Glimpses of
ANIMAL BIODIVERSITY

The Editor

Renowned environmentalist and Vice-Chancellor of Periyar University, Salem **Prof. Dr. K. Muthuchelian** has been working with dedication since 1982 for the development of India by harnessing both traditional and frontier technologies. He earned his M.Sc. (Botany) from Madurai Kamaraj University with distinction and Ph.D. from the School of Biological Sciences, Madurai Kamaraj University, Madurai, India and then he did his Post-Doctoral research at the University of Ancona, Italy. Dr.K. Muthuchelian was a Faculty member of the School of Energy, Environment and Natural Resources, Madurai Kamaraj University, Madurai. He was conferred the prestigious *"Doctor of Science"* (D.Sc.) in recognition of his research accomplishment on biomass technology. He became the Director of the *'Centre for Biodiversity and Forest Studies'* in Madurai Kamaraj University in 2001. His laboratory has got International Recognition (Nitrogen Fixing Tree Associations, Hawaii, USA) for 'Erythrina Research'.

He served as a member in various prestigious professional bodies such as The New York Academy of Sciences - USA, Rural Development Forestry Network – UK, American Association for the Advancement of Science - USA and International Union for Conservation of Nature (IUCN) and Natural Resources, Switzerland (an affiliated body of UNO) for Southeast Asia. He has been an expert member of *'Man and Biosphere (MAB) programme'* of Ministry of Environment and Forests, Government of India, New Delhi, India. He has been nominated as *"Fellow of International Energy Foundation, Saskatchewan, Canada"* in recognition of his outstanding International contributions to the 'Biomass production and energy transfer technology'. He has been elected as *"Fellow of National Academy of Biological Sciences, India"* for his excellent contribution to the 'Environmental Sciences'. He also served as a member and Chairman of the Peer Team of NAAC (National Assessment and Accreditation Council, Bangalore) in many premier academic institutions in India.

Dr. Muthuchelian's scientific contribution has been recognized through numerous awards such as *'Tamil Nadu Best Scientist in Environmental Sciences (1999)'* by TNSCST, Government of Tamil Nadu, Chennai, India, *'Best Scientist in Environmental Sciences (2001-2002)'* by Nehru Yuvakendra, Government of India and Tamil Nadu Sports Authority, *'Best Scientist in Environmental Management (2005) (Karma Veerar Kamarajar Award)'* by Department of Environment, Government of Tamil Nadu, Chennai, India. He has been awarded the *'Best Teacher Gold Medal (First Prize)'* by the Madurai Kamaraj University, Madurai in 2008. He was awarded the prestigious *'76th Indian Science Congress Endowment – Eminent Scientist Award in Natural Sciences'* by Madurai Kamaraj University in 2010. He has been honored twice the *"Merit of Excellence Award"* in recognition of his outstanding contributions in the field of medicinal plants conservation and in plant science at international level. For his remarkable contribution to the society and dedicated involvement in the enlistment of the downtrodden, he has been awarded twice *'Best Vice-Chancellor Award'* by the Indian Red Cross Society, Tamil Nadu branch, Chennai.

Dr. K. Muthuchelian is the First Indian Scientist honored with the prestigious award of the University of Ancona, Italy for his outstanding contribution in 'Bioenergy production'. He was an expert invitee of Nitrogen Fixing Tree Association, Hawaii, USA. He has attended several International Conferences and Symposia in USA, Germany, Hungary, Bangladesh and Italy. He has published more than 150 research papers in refereed National and International Journals and presented many papers in National and International Symposia/Conferences. He has organized 32 National Scientific Meetings, Workshops, Seminars and Conferences. Recently, he has published a book in Tamil entitled *"Uyir Virimam"* (Biodiversity: Current Status and Management) widely appreciated in both the academic and other professionals. He has produced 26 Ph.D's and currently guiding 12 Ph.D. scholars.

While ensuring accelerated environmental development through applications of science and technology, Dr. K. Muthuchelian has been relentlessly focusing attention on ecological conservation and on sustainable development. The integrated approach has remained one of the distinguishing aspects of Dr. K. Muthuchelian's original and path breaking contributions towards the revolutionary growth for ecological security and environmental sustainability in India and indeed the world.

Glimpses of
ANIMAL BIODIVERSITY

Editor

Dr. K. Muthuchelian
Ph.D., D.Sc., FNABS., FIEF (Canada)
Vice Chancellor
Periyar University
Periyar Palkalai Nagar
Salem – 636 011

2013
Daya Publishing House®
A Division of
Astral International Pvt. Ltd.
New Delhi – 110 002

Published by : **Daya Publishing House®**
A Division of
Astral International Pvt. Ltd.
– ISO 9001:2008 Certified Company
4760-61/23, Ansari Road, Darya
Ganj, New Delhi-110 002
Ph. 011-43549197, 23278134 E-mail:
info@astralint.com Website:
www.astralint.com

Laser Typesetting : **Classic Computer Services**
Delhi - 110 035

Printed at : **Salasar Imaging Systems**
Delhi - 110 035

PRINTED IN INDIA

Foreword

India has a rich and varied heritage of biodiversity covering ten biogeographical zones, the trans-Himalayan, the Himalayan, the Indian desert, the semi-arid zone(s), the Western Ghats, the Deccan Peninsula, the Gangetic Plain, North-East India, and the islands and coasts. India is rich at all levels of biodiversity and is one of the 17 mega diversity countries in the world. India's wide range of climatic and topographical features has resulted in a high level of ecosystem diversity encompassing forests, wetlands, grasslands, deserts, coastal and marine ecosystems, each with a unique assemblage of species.

Survey conducted so far in India have inventoried over 47,000 species of plants and over 89,000 species of animals over just 70 per cent of the country's total area. India's biogeographical location at the junction of the, Agrotrophical, Indo-Malayan and Paleo-Arctic realms has contributed to the biological richness of the country. The endemism of Indian biodiversity is high-about 33 per cent of the country's recorded flora is endemic to the country and is concentrated mainly in the North-East, Western Ghats, North- West Himalaya and the Andaman and Nicobar islands. About 62 per cent of the known amphibian species and 50 per cent of the lizards are endemic to India, the majority occurring in the Western Ghats.

The country is bestowed with immense agro-biodiversity and a rich diversity in landraces/traditional cultivars/farmer's varieties. A number of crop plants (384) are reported to be cultivated in India. A total of 49 indigenous major and minor crops have been reported in the "History of Agriculture in India" which include 5 cereals and minor millets, 4 pulses, 1 oilseed crop, 9 vegetables, 5 tuber crops, 11 fruits, 5 spices, 1 sugar yielding plant and 7 fiber crops. India is the centre of origin of 30,000-50,000 varieties of cultivated plants including rice, pigeon pea, mango, okra, bamboo etc.

The broad vision for biodiversity in Agenda 21st its conservation and sustainable use accompanied by equitable benefit sharing mechanism. This includes a focus on enhancing national biodiversity protection measures involving the development of national strategies; mainstreaming of biodiversity concerns; ensuring the fair and equitable sharing of the benefits accruing from biodiversity; country-wide studies on biodiversity; fostering traditional methods and indigenous knowledge; encouraging biotechnological innovations along with the suitable sharing of their benefits and promoting regional and international cooperation.

In a major advancement for the cause of biodiversity conservation in the country and in compliance with requirement of the Convention on Biological Diversity, the drafting of the country's National Biodiversity Strategy and Action Plan (NBSAP) with funding support from GEP, the Global Environmental Facility, is now underway. The strategy and action plan are very broad in scope and comprehensive in coverage and propose to prepare detailed action plans at sub-state, state, regional and national levels based on the framework Policy and Action Strategy on Biodiversity.

This book will bring out the first hand informations about diversity and bioresources both in land and water, and to manage the structured *in-situ* and *ex-situ* conservation measures.

I deeply indebted to thank all the contributors of this book for ready response within a short notice.

Dr. K. Muthuchelian

Preface

India is considered as one of the mega-biodiverse countries of the world. It has a bewildering diversity of flora and fauna, with a high degree of endemism especially Ampibian and Reptilian diversity which is one of the best in the world. Our inventory of invertebrate diversity is very high with only 10 per cent of the species being documented till now. With the increased pressure on wildlife in the form of climate change, habitat destruction, pollution, over exploitation the survival of animals have been at stake.

This book addresses the economic, institutional and social challenges confronting scientists and policy makers in conserving biodiversity and ecosystem services that are critical for sustaining human well being and development. The contributors to this book are experts who have made significant contribution to biodiversity research. The volume encompasses a wide range of themes and issues such as invertebrate biodiversity, vertebrate diversity, insect plant interaction, plankton diversity, climate change etc. The book includes chapters with focus from various climatic zones of India covering diverse ecosystems like tropical evergreen forests, tropical deciduous forest, agroecosystems, arid, freshwater, estuarine, marine ecosystems. Apart from these topics it also includes the faunal biodiversity ranging from zooplanktons, insects, fishes to large mammals in this changing environmental scenario.

In addition, we hope it will help natural resource managers, scientists and decision makers in overcoming their fear of models and help them in translating the model results into pro-active implementation to mitigate biodiversity loss.

This book also provides an updated information about the current state of animal distribution across India covering a variety of habitats. The book will provide useful information to students, environmentalists, foresters as well to the general public at large.

My heartfelt thanks are due to all contributors who made this compilation possible as well as my research students tremendously helped me in shaping this book.

Dr. K. Muthuchelian

Contents

2013, Glimpses of Animal Biodiversity   *Pages* **1–18**
Editor: **Dr. K. Muthuchelian,** *Vice Chancellor, Periyar University, Salem*
Published by: **Daya Publishing House, NEW DELHI**

Chapter 1

Faunal Diversity in India: An Overview

*P. Thangavel[1], G. Sridevi[2] and K. Muthuchelian[3] **

Department of Environmental Science, [3]Vice-Chancellor,
Periyar University, Periyar Palkalai Nagar,
Salem – 636 011, Tamil Nadu
[2]Department of Plant Biotechnology, School of Biotechnology,
Madurai Kamaraj University, Madurai – 625 021, Tamil Nadu

India: A Megadiversity Nation

India being one of the 17 megadiversity nation in the world with unique biogeographical location, diversified climatic conditions has been known for its rich biodiversity. India possesses a diversity of ecological habitats ranging from tropical, sub-tropical, alpine to desert. Biogeographically, India represents the two of the major realms (the Paleo-arctic and Indo-Malayan), five biomes [tropical humid forests, tropical dry/deciduous forests (including monsoon forests), warm/semi-deserts; coniferous forests and alpine meadows], ten biogeographic zones (Trans-Himalayan, Himalayan, Indian desert, Semi-arid, Western Ghats, Deccan Peninsula, Gangetic Plain, Northeast India, Islands and Coasts) and twenty-seven biogeographic provinces. The Indian landmass is bounded by the Himalayas in the north, Bay of Bengal in the east and the Arabian sea in the west and the Indian Ocean in the south with a total geographical area of about 3,029 million hectares. Miller Meier stated that, 'India is remarkable in both species richness and endemism although it ranks 10th position' (Sinha *et al.,* 2010).

---

* Corresponding Author

In India, the Western Ghat forests and the Eastern Himalayan forests have been identified as the megadiversity hot spots. In fact, within the 2.4 per cent of the world's total land area, India is known to have 11 per cent and 7.5 per cent of the flora and fauna, respectively. The Zoological Survey of India (ZSI) established in 1916, have surveyed and recorded nearly 92,037 species so far. Nearly 80 per cent are insects which include 61,375 species (Table 1.1; ZSI, 2012). Nearly 140 breeds of native domesticated animals such as cattle (30), buffalo (10), sheep (42), goat (20), camel (9), horse (6), donkey (2) and poultry (18) were also found (NBAP, 2008). Amphibians constitute about 62 per cent and majority of them are found in Western Ghats and 32

Table 1.1: Estimated Faunal Diversity in India

Taxonomic Group	No. of Species		Per cent in India
	World	India	
PROTISTA (Protozoa)	31250	2577	8.24
ANIMALIA			
Mesozoa	71	10	14.08
Porifera	4562	500	10.70
Cnidaria	9923	999	10.07
Ctenophora	100	12	12.00
Platyhelminthes	17511	1639	9.35
Rotifera	2500	330	13.20
Gastritricha	3000	100	3.33
Kinorhyncha	100	10	10.00
Nematoda	30028	2878	9.58
Acanthocephala	800	229	28.62
Sipuncula	145	35	24.14
Mollusca	66535	5155	7.75
Echiura	127	43	33.86
Annelida	12701	842	6.63
Onychophora	100	1	1.00
Arthropoda	999059	71480	7.15
Crustacea	35536	2941	8.28
Insecta	867516	61375	7.07
Arachnida	73444	5833	7.94
Pycnogonida	600	17	2.83
Chilopoda	3000	100	3.33
Diplopoda	7500	162	2.16
Symphyla	120	4	3.33
Merostomata	4	2	50.00
Phoronida	11	3	27.27

Contd...

Table 1.1–*Contd...*

Taxonomic Group	No. of Species		Per cent in India
	World	*India*	
Bryozoa (Ectoprocta)	4000	200	5.00
Entoprocta	60	10	16.66
Brachiopoda	300	3	1.00
Chaetognatha	111	30	27.02
Tardigrada	514	30	5.83
Echinodermata	6223	767	12.33
Hemichordata	120	12	10.00
Chordata	46499	5163	10.65
Protochordata	2106	119	5.65
Pisces	21734	2641	12.15
Amphibia	5185	312	6.02
Reptilia	5819	462	7.94
Aves	9026	1232	13.66
Mammalia	4629	397	8.58
Total (Animalia)	1195759	89460	7.48
Grand Total (Protista + Animalia)	**1227009**	**92037**	**7.50**

Source: Animal Discovery 2011, Compiled by: Director, ZSI (Updated January, 2012).

per cent of the reptiles are endemic. Nearly one third of the animal varieties found in India were reported in Western Ghats of Kerala alone. The remaining species which is estimated about twice the number of the species are yet to be discovered.

The year of 2010 is considered as the "*International Year of Biodiversity*" which includes a series of programme to protect the biodiversity. India is hosting the Eleventh Conference of Parties (CoP) to the Convention on Biodiversity (CBD) in Hyderabad on 1-9 October, 2012. India with a strong commitment is contributing towards achievement of three objectives of the CBD, the 2010 target and the strategic plan. The CBD will provide India an opportunity to consolidate, scale-up and demonstrate our initiatives and strengths on biodiversity. The Ministry of Environment and Forests (MoEF) framed and formulated the strategy for conservation and sustainable utilization of biodiversity in India. With a strong institutional, legal and policy framework, India has the potential and capability to emerge as the world leader in conservation and sustainable use of biodiversity. The MoEF has initiated a major new programme in collaboration with the Economics of Ecosystems and Biodiversity (TEEB) to evaluate the wealth of natural resources and biodiversity in India in terms of economic value. During the year 2011-2012 survey, 19,010 examples of various groups of animals belonging to 793 species were identified by the scientists of ZSI. Additionally, 29 new species have been identified and 15 species were added as new fauna of India (MoEF, 2012).

The prime objective of this chapter is to highlight the faunal resources of India which will provide an extremely valuable source of information. This chapter also review the important faunal diversity in India based on the literature survey of existing information combined with key information provided by various National and International organizations, ministries and research institutions in India.

Endemic Faunal Species in India

India possesses a diverse range of endemism across the faunal groups. The north-east India, the Western Ghats and the north-western and eastern Himalayas are the major hot spot areas rich in endemism. Although a small pocket of local endemism occurs in Eastern Ghats, the Gangetic plains possess poor endemics. The lower groups such as Mesozoa (100 per cent), Oligochaeta (77.8 per cent), Platyhelminthes (71.9 per cent) and Kinorhyncha (70 per cent) showed very high endemism. The endemism of the Indian reptiles (187 reptiles) and amphibians (110 species) are high among the higher groups. Around eight amphibian genera are not found outside India.

A relatively low endemism exists among mammals and birds. In Western Ghats, the four endemic species of conservation significance includes the Lion-tailed macaque *Macaca silenus*, Nilgiri leaf monkey *Trachypithecus johni*, Brown palm civet *Paradoxurus jerdoni* and Nilgiri tahr *Hemitragus hylocrius*. Among the birds, only 55 species are endemic in India and their distribution is concentrated in the high rainfall regions such as eastern India along the mountains and monsoon shadow regions of south-west India and the Andaman and Nicobar islands (Sinha *et al.*, 2010). Of the known 140 bird species, the Great Indian Bustard is a globally threatened species. Nearly, 38.80 per cent of the India's total geographical area is covered by arid and semi-arid regions. The Trans-Himalayan region located in the cold arid zone covers about 5.62 per cent of the total area of the country. The major three cat predators – the lion, leopard and the tiger is the major stronghold in this region. The cold desert is the home for endangered faunal species such as *Asiatic ibex*, Tibetan argali, Wild yak, Snow leopard etc. As per the International Union for Conservation of Nature (IUCN), there are about 413 globally threatened faunal species in India which accounts for 4.9 per cent of world's total number of threatened faunal species (IUCN, 2008). India is the third largest producer of fish in the world with vast inland and marine bioresources. The database of fishes found in India includes 2,182 fishes among which 327 freshwater species listed in IUCN were considered under threat categories and nearly 192 were endemic fishes. Similarly, there are two endemic families and 127 monotypic genera of fish fauna in India (Goyal and Arora, 2009).

Status of Important Indian Faunal Diversity

Fishes

Among the 22,000 species of fishes recorded worldwide, about 11 per cent are recorded in Indian waters. Globally, there are about 450 families of freshwater fishes of which 40 are roughly present in India alone and 25 of these families consist of commercially important species (Jaiswal and Ahirrao, 2012). Major warm water species include: *Bagarius bagarius, Catla catla, Channa marulius, C. punctatus, C. striatus,*

Cirrhinus mrigala, Clarias batrachus, Heteropneustes fossilis, Labeo bata, L. calbasu, L. rohita, Aorichthys seenghala, Notopterus chitala, N. notopterus, Pangasius pangasius, Rita rita and *Wallago attu.* There are about 1,367 fish species in Indian Ocean, 603 species of fishes in Lakshadweep Islands (Jones and Kumaran, 1980), more than 1,000 species in Andaman and Nicobar Islands and about 538 in Gulf of Mannar Biosphere Reserve (Venkataraman and Wafar, 2005). The National Bureau of Fish Genetic Resources (NBFGR) focuses on the development of fish databases, genetic characterization, fish germplasm and habitat inventory, risk analysis of exotic species and other areas of germplasm conservation, etc. The database includes 2,508 native finfishes reported in India belonging to 928 genera of 237 families under 39 orders. The distribution of Indian fishes based on their habitat is provided in the Table 1.2. Currently, the NBFGR website has records for 220 fish species belonging to 13 families (Tyagi and Jena, 2012). The NBFGR vision 2030 is geared to safeguard the aquatic biological wealth of the nation and also towards sustainable utilization for nutritional security of the masses (Lal and Jena, 2011).

Table 1.2: Fish diversity of India.

Category of Fishes	Ecosystem	No. of Fish Species
Native fishes	Freshwater	877
	Brackish water	113
	Marine water	1518
	Total	**2508**
Exotic fishes		291
Grand Total		**2799**

Source: National Bureau of Fish Genetic Resources, Annual Report 2011-12.

India has a rich fish germplasm resources which includes a large number of endemic species. The Indian fish fauna is classified into two classes: Chondrichthyes (represented by 131 species under 67 genera, 28 families and 10 Orders) and Osteichthyes (represented by 2,415 species belonging to 902 genera, 226 families and 30 orders). Among the five families of Indian Osteichthyes, the fishes of the family Parapsilorhynchidae with the only genus Psilorhynchus is endemic to India. Other endemic fishes include the genus Olytra (family Olyridae) and the species *Horaichthys setnai* (family Horaichthyidae). Among the total bony fish families of Indian origin, 2.21 per cent fish families are endemic. A total of 223 endemic fish species (8.75 per cent of total fish species) are found in India belonging to 128 monotypic genera of fishes representing 13.20 per cent of fish genera from Indian region. According to IUCN, there are 1,275 globally threatened fish species. In India, as per IUCN classification, about 42 Indian fish species are globally threatened (www.iucnredlist.org). The MoEF listed Pondicherry Shark (*Carcharhinus hemiodon*), Ganges Shark (*Glyphis gangeticus*), Knife-tooth Sawfish (*Anoxypristis cuspidate*), Large-tooth Sawfish (*Pristis microdon*) and Long-comb Sawfish or Narrow-snout Sawfish (*Pristis zijsron*) as critically endangered species in India (MoEF, 2011a).

Gharial

The only surviving member of Gavialidae family is the Gharial. The estimated population of wild Gharial across the globe was about 5,000 to 10,000 in the 1940s, which declined (almost 96 per cent) to less than 200 in 1976. India stood again at a poor figure of less than 200 individuals in 2006 (Andrews, 2006). In Indian sub-continent, Gharials (*Gavialis gangeticus*) are endemic and there have been a drastic decline in the Gharial population in Southern Asia (Maskey, 1999) in the last 30 years. Gharial has now become extinct in countries like Pakistan (Whitaker and Basu, 1983), Bangladesh (Whitaker, 1976; Khan, 1979; Faizuddin, 1985), Burma and Bhutan (Singh, 1991). The two rivers, Girwa and the Chambal are now confined to be the major breeding places for Gharials while Nepal has only a remnant breeding population. A combination of factors such as, habitat alteration, destruction, prey depletion, direct mortality, pollution, siltation and hunting for skin and medicinal purpose led to an irreversible loss of riverine habitat and consequently the Gharial. A survey in 1974 concluded that only 50-60 survived in India while a handful existed in Nepal, Pakistan and Bangladesh. MoEF has also constituted a National Tri-State-Chambal Sanctuary Management and Coordination Committee (NTRIS-CASMACC) for protection, conservation and recovery of critically endangered Gharial (*Gavialis gangeticus*) in its natural habitats in Chambal and Girwa rivers in three States of Madhya Pradesh, Rajasthan and Uttar Pradesh. In the IUCN Red list of endangered species, Gharial is uplisted as "Critically Endangered" and is listed in the Appendix I of CITES (Saikia *et al.,* 2010). The Wildlife Protection Act, 1972 clearly prohibited the hunting. An active programme *viz.,* "Project Crocodile" was launched successfully in India to conserve the Gharials in 1975 with the aid of the United Nations Development Fund (UNDP) and Food and Agricultural Organization (FAO). Between 1975 and 1982, sixteen crocodile rehabilitation centers and five crocodile sanctuaries were established in India (Whitaker, 2007). The current world population includes 220 wild reproducing Gharials. In India, the National Chambal Sanctuary holds about 90 per cent of the total population of Gharial which is followed by Katerniaghat Wildlife Sanctuary and Son Wildlife Sanctuary.

Bustard Species

In India, the four bustard species are in great trouble. India has the unfortunate distinction for the home of three rarest bustard species among the 23 bustard species in the world. As India is the 'custodian country' for three resident bustard species, if they disappear from India, they will become extinct from the world also. Two are critically endangered (the Bengal Florican and Great Indian Bustard) and one is endangered (the Lesser Florican) while the other is vulnerable (the Houbara Bustard, *Chlamydotis undulate*) as per the BirdLife International and IUCN criteria. Hence, there is a requirement of responsibility to protect these bustard species to halt their extinction (WWF-India, 2012).

(a) Great Indian Bustard

The Great Indian Bustard (GIB) (*Ardeotis nigriceps* also known as *Choriotis nigriceps*) exists in the short grass plains and desert plains of west Rajasthan and Gujarat. It has been declared as the State Bird of Rajasthan. Among the bustard family, Indian bustard is the most endangered member in the world and their total population may not exceed 700. Due to its small and declining population, the bird is on the 'Critically Endangered' red list of IUCN in 2011. The degradation of grasslands and conversion into other forms, habitat fragmentation or degradation due to invasive species, agriculture expansion and change of floral composition of grasslands and general increase in anthropogenic pressures are some of the major threats to the GIB. With the support from MoEF, WWF-India, Bombay History and Natural Society (BHNS) and many other organizations have developed recovery plans for bustard. The MoEF has sanctioned a project to Gujarat Ecological Education and Research Foundation (GEER) Foundation for monitoring GIB population in Gujarat. In addition, various conservation measures have been undertaken in order to ensure the conservation of GIB in the long run. As the globally threatened bird species are not present in Project Tiger areas and/or forests, they are not benefited by Project Tiger. Similar to Project Tiger, Project Elephant, and Project Snow Leopard, recovery plans like 'Project Bustard' should be implemented to save these species.

(b) Lesser Florican (*Sypheotides indica*)

The Lesser Florican moves to Gujarat, western Madhya Pradesh, north-western Maharashtra and south-eastern Rajasthan at the beginning of the monsoon for breeding purpose. Later, it disperses to south and east of the Indian subcontinent, where there is little clear evidence about its status, distribution or habitat requirements (Sankaran, 1993, 1997). This species is endemic to Indian subcontinent and it's in the 'Endangered' in IUCN Red List, 2011. The Integrated Development of Wildlife Habitats by the MoEF, 2009 included this as one of the priority species for the recovery programme.

(c) Bengal Florican (*Houbaropsis bengalensis*)

The Bengal Florican was widely distributed in India, Nepal and Bangladesh but currently restricted to a few protected areas in India [Uttar Pradesh (70-80), Assam (180-220) and Arunachal Pradesh (40-50)] (BirdLife International, 2010). The distribution of Bengal Florican is estimated to be less than 350 birds in India and it's 'Critically Endangered' in the IUCN Red List (IUCN, 2008). It is also listed under Schedule I of the WPA and in the National Wildlife Action Plan 2002-2016. The Integrated Development of Wildlife Habitats of the MoEF has also identified this as one of the species for recovery programme.

Vulture Crisis in India

There are about nine species of vultures found in India. They are: White-rumped Vulture, Long-billed Vulture, Slender-billed Vulture, Red-headed (king) Vulture, Egyptian Vulture, Himalayan Griffon, Eurasian Griffon, Cinereous Vulture and Lammergier Vulture. There has been a decline by about 95 per cent in the population of three Gyps vultures (white-rumped vulture, long-billed vulture, and slender-billed vulture) since early 1990s in the Indian sub-continent. All the three vulture species were listed by IUCN as 'Critically Endangered' in 2000, which is the highest category of endangerment. These assessments indicated a high risk of extinction in near future too. The catastrophic vulture crisis is mainly due to food scarcity, decline in the number of tall trees for nesting and roosting, and anthropogenic pressures, industrialization and urbanization. The study on the sudden decline of the Gyps vultures in the Indian subcontinent was first recorded at the Keoladeo Ghana National Park, Rajasthan in the mid 1980's – 1990's clearly indicated the use of diclofenac in veterinary applications that allow diclofenac to present in carcasses of domestic livestock is available as food for vultures. The decline has been projected over 97 per cent in India within 12 year period (Virani, 2006). There were about 2,647 Gyps vultures present in the state of Gujarat during 2005 as surveyed by GEER Foundation. In April 2006, the 'Action Plan for Vulture Conservation in India' was launched by the MoEF. Also, the 'IUCN South Asian Task Force' was established to implement national vulture recovery and also include conservation breeding and release. Various measures for conservation of vultures in India includes: establishment of Vulture Breeding Centres in Haryana, West Bengal and Gujarat; raising awareness among people, policymakers, and other stakeholders and also monitoring the vulture population. Recently, the Government of India has encouraged various projects for the identification of alternative drugs to diclofenac to save the vulture population (MoEF, 2006).

Sarus Crane

Among the 15 crane species found in the world, six of them are found in India. The Sarus Crane (*Grus antigone*), world's tallest flying bird, is the only resident species of India and is the state bird of Uttar Pradesh. The Sarus cranes have been reported in the states of Bihar, West Bengal, Assam, Rajasthan, Gujarat and Madhya Pradesh, but majority (2,500 – 3,000) are found in Uttar Pradesh. The Sarus cranes preferred to live in the same geographic area and non-migratory in nature. There are three sub species of Sarus Crane *viz.*, the Eastern Sarus Crane, the Australian Sarus Crane and the Indian Sarus Crane with a total estimated population between 13,500 and 15,500. The Sarus Cranes live in association with agricultural landscape and in harmony with farmers (Gole, 1989; Archibald and Meine, 1996) in regions like wetlands, flood areas, ponds, lakes, rice paddies, etc. The Indian Sarus Crane, although common in northern India due to high population pressures continues to decline where it still exists. The most significant threats to the species in India includes: degradation of wetlands, pollution, heavy use of pesticides and industrial development. Other threats include high-tension electric cables, poaching for meat, egg, pet trade, and disturbance by the farmers during the nesting period in the field. The Sarus is classified as 'Endangered' in the revised IUCN Red List Categories.

Sparrow

The house sparrow (*Passer domesticus*) are relatively sedentary birds and don't travel more than one or two kilometres in search of food. Nearly 70 per cent of the sparrow population has been declined now days in certain places in India. The population of sparrows declined 80 per cent in Andhra Pradesh, 20 per cent in Kerala, Gujarat and Rajasthan whereas in coastal areas there was a sharp decline of 70-80 per cent (Dandapat *et al.,* 2010). The Ministry of Environment and Forests (MoEF), Government of India constituted an expert committee to evaluate and study the possible impacts of Communication Towers on Wildlife including birds and bees on August 30, 2010. The review committee reported that the electro-magnetic radiations (EMR) did interfere with the biological systems and may have varying impacts on wildlife especially birds and bees. Based on the report of the committee, necessary actions were undertaken by various agencies to avoid and mitigate the impacts of EMR.

Tiger

The India's national animal, the Tiger is a symbol that is an intrinsic part of our culture. One of the earliest portrayals of the tiger in India dated back to 2500 BC in the Harappan seals from the Indus valley culture. The Tiger Task Force was initiated by Mrs. Indira Gandhi under the chairmanship of Dr. Karan Singh in 1970 for tiger conservation with an ecological approach. This laid foundation for the emergence of India's tiger conservation programme – Project Tiger in 1972 (http://projecttiger.nic.in). The Project Tiger has been spread out in 17 states with 40 tiger reserves accounting to almost 1 per cent of the country's geographical area. The new Tiger Reserve, Kawal created in Andhra Pradesh has been approved by the National Tiger Conservation Authority (NTCA). The approval of other five tiger reserves such as Pilibhit (Uttar Pradesh), Ratapani (Madhya Pradesh), Sunabeda (Odisha), Mukundara Hills (including Darrah, Jawahar Sagar and Chambal Wildlife Sanctuaries) (Rajasthan) and Kudremukh (Karnataka) has been accorded by NTCA (MoEF, 2011a). The proposals have been initiated for declaring the following areas as Tiger Reserves:

1. Bor (Maharashtra),
2. Suhelwa (Uttar Pradesh),
3. Nagzira-Navegaon (Maharashtra),
4. Satyamangalam (Tamil Nadu),
5. Guru Ghasidas National Park (Chhattisgarh),
6. Mhadei Sanctuary (Goa) and
7. Srivilliputhur Grizzled Giant Squirrel/Megamalai Wildlife Sanctuaries/ Varushanadu Valley (Tamil Nadu).

According to the data collected in 2006, there are about 1,411 tigers in India and the census in July 2011 reported about 1,706 tigers in 39 Tiger Reserves. Among the 39 tiger reserves, 15 were rated as 'very good', 12 as 'good', 8 as 'satisfactory' and 4 as 'poor'. The project started with the full assistance initially from the Central Government till 1979-1980, which later become a Centrally Sponsored Scheme. During the XI plan period, assistance was provided to the ongoing Centrally Sponsored Scheme of Project Tiger from Rs. 650 crores to Rs. 1,216.86 (MoEF, 2011a, b).

In the last century, various pressures such as the development of wireless communication system and outstation patrol camps have declined the tiger population considerably. The most iconic species of Asia, the tiger have decreased in the wild from 100,000 individuals to an estimate of 3,200 to 3,500 (Global Tiger Initiative, 2011), with a decline in Asia's forests by more than 70 per cent in half of the countries in this region (Laurance, 2007). According to The Living Forests Model, the available habitats for tiger will shrink by 2050 based on the projections on past trend. This projection does not include the steps taken by national and local policies to protect forest resources (Strassburg *et al.*, 2012). The Project Tiger has saved the tiger from extinction by improving the protection and status of its habitat. Various developmental works have been launched to improve the water regime, ground and field level vegetations, control of live stock grazing etc. to increase the animal density. Overall, the restoration management and intensive protection under the Project Tiger have saved many of our eco-typical areas from destruction. A Global Tiger Forum for Tiger Range Countries have been created for addressing the international issues related to tiger conservation. An active management is being carried out in the Sariska and Panna tiger reserves in India to reintroduce tigers where tigers have become locally extinct (MoEF, 2011a).

Elephant

The largest terrestrial mammal of India is elephant (*Elephas maximus*) and requires larger area. The status of elephants serves as the best indicator to determine the forests status. They are widely distributed from Tigris – Euphrates in West Asia eastward through Persia in to the sub-continent India, South and Southeast Asia. Presently, the elephant population is confined to Indian subcontinent, Southeast Asia and some Asian Islands – SriLanka, Indonesia and Malaysia. In India, they are distributed mainly in South India, Northeast including North West Bengal, Central Indian states of Odisha, Southwest Bengal, Jharkhand, and Northwest India in Uttarkhand and Uttar Pradesh. The Government of India launched the Project Elephant (PE) in the year 1992 to protect the elephants, their habitat and corridors, to address the man-animal conflict and also the welfare of domesticated elephants. There are about 28 Elephant Reserves (ERs) which cover an area of about 60,000 sq. km. as per the notification of various State Governments. Yet the consent for establishment of 6 more ERs in Baitarini and South Odisha ER (Orissa); Lemru and Badalkhod (Chattisgarh); Ganga-Jamuna (Shiwalik) ER in Uttar Pradesh; Khasi ER (Meghalaya) is under consideration in MoEF. The population estimates of elephants were reported to be around 25,604 in 1993, 25,877 in 1997, 26,413 in 2002 and 27,694 in 2007-08 (Rangarajan *et al.,* 2010). The above scenario clearly reflects an increase of elephant population over the past 15 years.

The elephant (Gajah) and the people (Praja) have to go together for successful survival of the elephant. A serious attention and requires immediate action is the Human-elephant conflict. Every year 400 people die due to elephants and in turn half of the 100 elephants were killed each year as defense to protect their crops. It's tragic both for elephants and humans as both are victims of victims. The important facets of elephant conservation include crop compensation and ex-gratia payment for the human life loss. A large number of endangered wild animals including elephants, tigers, leopards, rhinoceros and gaur are killed by train hit also. The country has lost 150 elephants due to train hits since 1987 from various states - 36 per cent (Assam), 10 per cent (Jharkhand), 6 per cent (Tamil Nadu), 3 per cent (Uttar Pradesh), 3 per cent (Kerala) and 2 per cent (Odisha). Various factors such as movement of elephants for food, water, shelter, and vegetation and also speed of train, the lack of awareness among the drivers, passengers, etc., contribute towards the elephant mortality by train hits. In addition, the most critical factor for the decline of elephants is the illegal poaching of male elephants for ivory trade (Rangarajan *et al.,* 2010).

In the year 2003, Monitoring of Illegal Killing of Elephants (MIKE) program was started in South Asia to build institutional capacity for long-term management of elephant populations within the range States. Also, they measure the levels and trends in the illegal hunting of elephants, the changes in their trend over time and to determine the factors or associated changes and to assess the trends of decisions taken by the Conference of the Parties to CITES. The following are the MIKE sites in India: Chirang Ripu, Dhang Patki (Assam); Eastern Dooars (Arunachal Pradesh); Garo Hills (Meghalaya); Mayurbhanj (Odisha); Mysore (Karnataka); Nilgiri (Tamil Nadu); Shivalik (Uttarakhand) and Wayanad (Kerala). The data are collected periodically every month by MIKE patrol and submitted to Sub Regional Support Office for South Asia Programme located in Delhi who assists the Ministry for the implementation of this programme. Hence, the declaring the elephant as National Heritage Animal will recognize its dual identity as an emblem of ecological sensitivity and of culture.

Great Indian One-Horned Rhinoceros

The Great one-horned rhino is commonly present in Nepal, Bhutan, Pakistan and in India primarily in northern India and Assam. As their Latin name, *Rhinoceros unicornis* suggests, the Indian Rhinos have only one horn that can be distinguished from other rhinos. The alluvial flood plains and tall grassland areas along the foothills of Himalayas are the preferred habitat for Indian Rhinoceros.

The prominent horn in these Rhinos is known well for their downfall. Their hard, hair-like growth is used for medicinal purpose in China, Taiwan, Hong Kong and Singapore and this is the main reason for their decline. The horn is valued as an ornamental dagger handle in North Africa and Middle East. Over the years, the destruction of their habitat has brought the rhinos to the brink of extinction. In the world, no more than 2,000 remain while only two populations contain more than 100 rhinos – Kaziranga National Park in Assam, India (1,200) and Chitwan National Park (CNP), Nepal (600). In spite of the joint efforts by India and Bhutan, the survival of a small population of rhinos remains doubtful along the Indo-Bhutan border. The Indian and Nepalese Governments with the help of the World Wildlife Fund (WWF) have taken large measures towards the conservation of Rhinoceros. In India, the Kaziranga National Park, Manas National Park, Pabitora Reserve Forest (highest density of Indian rhinos in the world), Orang National Park, and Laokhowa Reserve forest in Assam and Royal Chitwan National Park, Nepal serve as the homes for this endangered animal. The conservation of one-horned rhino has been a great success in Assam and India. The number of species declined to 10-20 rhinos in Kaziranga National Park, Assam in 1905. The population has been recovered to over 1700 individuals upon strict protection. Today, about 2,400 only survive in India and Nepal. The Indian Rhino Vision 2020 (IRV) is being implemented by Department of Environment and Forests, Assam Government to increase the total rhino population to 3000 by the year 2020. The first phase of the programme has already been launched from July 2005 onwards for three years until June 2008. This included the improved protection of rhinos and translocation of 20-30 rhinos from Pabitora and Kaziranga to Manas National Park where they are protected and monitored (IRV 2020).

Snow Leopard

Nearly, 7500 snow leopard (*Panthera pardus*) are estimated to survive in the Himalayas and Central Asian mountains. Atleast 10 per cent of the global population is found in India. Very little is known about the snow leopard (McCarthy and Chapron,

2003), which is now a globally endangered species as per the IUCN's Red List. The Project Snow Leopard was launched by Indian Government for strengthening wildlife conservation in the Himalayan high altitudes. Leopards are fully protected under the domestic legislation of India and commercial trade is banned under CITES. During 2001-2010, the study documents a total of 420 seizures of leopard skins, bones and other body parts reported from 209 localities in 21 out of 35 territories in India. Uttarakhand emerged as a major source for trading the leopard parts while Delhi served as a major epicenter of the illegal trade along with the adjacent states. Mostly, the leopard parts are smuggled out of India to other countries in Asia. An official database "Tigernet" (http://www.tigernet.nic.in/) used for Tiger conservation in India would also help in monitoring the illegal trade of leopard parts.

Kashmir Hangul Deer

Hangul, an endangered species of red deer is found only in Kashmir. The CITES listed the Hangul (*Cervus elaphus hanglu*) as a critically endangered species of deer and it is the only survivor of red deer in the Indian sub-continent. The deer population had shown a declining trend from 1940s to 1989, the numbers fell to 900 from 3000. According the latest census in 2011, the number of Hangul recorded in Dachigam National Park was 218 (Hussain, 2012). Further, another research programme has been launched outside the Park to study the relic population of Hangul in collaboration with the Wildlife Trust of India.

Ganges River Dolphin

Ganges River Dolphin was declared India's National Aquatic Animal in 2009. The Ganges River Dolphin, or susu, inhabits Ganges-Brahmaputra-Meghna and Karnaphuli-Sangu river systems of Nepal, India, and Bangladesh. The Ganges River Dolphin is one of four freshwater dolphins of the world. The river dolphin is often known as the 'Tiger of the Ganges' because it occupies the same position in a river ecosystem as that of a tiger in a forest. The population of Ganges River Dolphin has declined from 6,000 in 1982 to less than 1,800 in 2005 due to construction of dams and water pollution caused by various anthropogenic sources and industrial effluents. According to the sources from WWF-India, the annual mortality rate of this animal was estimated to be 130-160 (Deccan Herald, 2012). The mammal is listed in the Schedule I of the Indian Wildlife (Protection) Act, 1972 and categorised as "endangered" by the World Conservation Union. Recently, the Uttar Pradesh government has launched a three day awareness programme namely "My Ganga, My Dolphin" during October 5-7, 2012. This campaign also surveyed the number of Gangetic River Dolphins present across a 2800 km stretch of the River Ganga and its tributaries including Yamuna, Son, Ken, Betwa, Ghagra and Geruwa.

Biodiversity Hotspots in India

Northeastern India and Eastern Himalayas

The Eastern Himalayas, sites at the biogeographical crossroads of two continental plates, possesses an incredible biodiversity wealth which render them a global biodiversity hotspot. The Eastern Himalayas, is the home for 300 mammal species,

977 bird species, 176 reptiles, 105 amphibians, and 269 freshwater fish. This region harbours a high density of the Bengal tiger and the presence of one-horned rhinoceros. Atleast 353 new species have been discovered in Eastern Himalayas between 1998 and 2008, which accounts to an average of 35 new species every year in the past ten years (Thompson, 2009). The discoveries include 242 plants, 16 amphibians, 16 reptiles, 14 fish, two birds and two mammals and atleast 61 new vertebrates. Most of the new amphibians discovered in Eastern Himalayas are endemic. The flying frog, *Rhacophorus suffry* was described in 2007 originally found in Assam. Other species are Assamese cascade frog, *Amolops assamensis* and caecilian, *Ichthyophis garoensis* from Assam and *Philautus sahai* described in 2006 from Arunachal Pradesh. The 16 new reptiles reported include 13 lizards and three snakes. Sikkim is known for its richness in avian diversity, which is further enriched by the presence of over 627 species of butterflies and insects (Sikkim Biodiversity Action Plan). Leaf Deer (*Muntiacus putaoensis*), a new species of barking deer discovered in Myanmar, was reported in the forests of Arunachal Pradesh in 2003 (Datta *et al.*, 2003). Of the 15 known species of three families of primates, nine species occur in Northeast India. The Golden Langur (*Trachypithecus geei*) is localized in the foothills of Himalays and this species is narrow endemic. The Stump-tailed Macaque (*Macaca arctoides*) and Northern Pigtailed Macaque (*M. leonine*), sympatrically distributed in Northeast India have become endangered (Chatterjee *et al.*, 2006).

Among the cat family, the Clouded Leopard (*Neofelis nebulosa*) is restricted to Northeastern region. The other members of cat family, the tiger, leopard, snow leopard along with clouded leopard were sustained in Arunachal Pradesh. Red Panda, listed as 'Endangered' in IUCN, habitat predominantly (23,000 sq. km) in Arunachal Pradesh (Choudhury, 2002). All the bear species listed in India are recorded in the northeast region. About 33 per cent of the elephant population of India is found in northeast India, especially in the north bank of Brahmaputra river in Assam. Both species of Great Indian Rhinoceros, One-horned Javan Rhinoceros (*Rhinoceros sondaicus*) and two-horned Sumatran Rhinoceros (*Didermocerus sumatrensis*) found in northeast India are now extinct from the region (Chatterjee *et al.*, 2006). The Brow-antlered Deer (*Cervus eldi eldi*), one of the rarest and most localized subspecies of deer in the world is endemic to Manipur. Among the seven endemic bird areas in the country, the eastern Himalayas and Assam plains are two endemic areas with highest diversity of bird species. Nearly, 760 bird species have been reported from Arunachal Pradesh (Borang, 2004) in which the Elliot's Laughing Thrush (*Garrulax elliotii*) and Brown-cheeked Laughing Thrush (*G. henrici*) were added newly to the list. The Whitewinged Duck, reported from D'Ering Wildlife Santuary of Arunachal Pradesh is highly endangered (Kaul, 2000). The Black Necked Crane is habitat only in the Sangte Valley of Arunachal Pradesh.

Few of the globally threatened birds such as Spot-billed Pelican (*Pelicanus philippensis*), Blacknecked Stork (*Ephippiorhyncus asiaticus*), Lesser Adjutant (*Leptotilos javanicus*) and Pale-capped pigeon (*Columba punicea*) are now found in Assam. The highly endangered Rufous-vented Prinia, is reported from the Pobiotora Wildlife Sanctuary in Assam. The beautiful Nuthatch (*Sita formosa*) is a resident in the forests of northeast India (Chatterjee *et al.*, 2006). In northeast India, so far 137 species of reptiles are recorded and Gharials from Brahmaputra is of greater conservation significance. Also, the northeast India possesses highest diversity of turtles. The black softshell turtle (*Aspideretes nigricans*), which was considered as extinct at one time has been rediscovered in Assam valley. Further, the Assam state also consists of 20 lizard species and 58 species of snakes, while the other tiny states include 18 lizard species and 34 snakes from Manipur and 92 snakes from Arunachal Pradesh. King Cobra (*Ophiophagus Hannah*) is the most awe-inspiring reptile in Arunachal Pradesh region. *Typhlops jerdoni, T. tenuicollis, Stoliczkaia khasiensis, Elaphe mandarina, Oligodon melazonotus, Xenochrophis punctulatus, Bungarus bungaroides, Trimeresurus jerdoni* are just a few examples of very elusive and rare snakes of Northeast India (Chatterjee *et al.*, 2006). Nearly, 64 amphibian species have been recorded in north east India. The state of Nagaland recorded 19 new amphibian species between 1998 and 2002 of which 5 species are new records for India. The Orang Sticky Frog (*Kalophrynus orangensis*) is present only in the Orang Wildlife Sanctuary of Assam (Chatterjee *et al.*, 2006).

There were about 3,624 species of insects and 50 molluscs recorded in the northeast region and they contribute maximum number of butterflies and moths in the country. Nearly, 350 species of butterflies were recorded in altitudes less than 900 m and the family Nymphalidae forms 50 per cent of the observed species followed by Lycaenidae and Pieridae (each 17.2 per cent). The families, Papilionidae and Hesperiidae are relatively low species richness contributing 8.65 and 7.0 per cent of the species. The largest tropical Lepidoptera, Atlas Moth (*Attacus atlas*) is not uncommon in many parts of the northeastern India. The most beautiful butterflies in the country, *Princeps polyctor ganesa*, and most beautiful moths, *Erysmia pulchella* and *Nyctalemon patroclus* is present in the northeast region (Chatterjee *et al.*, 2006).

Western Ghats

The hill ranges of the Western Ghats, a global biodiversity hotspot, extend along the west coast of India from the river Tapti in the north to the southern tip of India. Their positioning makes the Western Ghats biologically rich and biogeographically unique - a veritable treasure house of biodiversity. Though covering an area of 180,000 square kilometres, or just under 6 per cent of the land area of India, the Western Ghats contain more than 30 per cent of all plant, fish, herpeto-fauna, bird, and mammal species found in India. Many species are endemic, such as the Nilgiri tahr *(Hemitragus hylocrius)* and the Lion-tailed macaque *(Macaca silenus)*, in fact 50 per cent of India's amphibians and 67 per cent of fish species are endemic to this region. The region has a spectacular assemblage of large mammals - around 30 per cent of the world's Asian elephant *(Elephas maximus)* population and 17 per cent of the world's existing tigers *(Panthera tigris)* call this area their home.

There are 332 species of butterflies (36 endemic), 290 species of fishes (116 endemic), 181 species of amphibians (94 endemic), 203 species of reptiles (124 endemic), 508 species of birds (19 endemic) and 137 species of mammals (16 endemic) are reported from Western Ghats. Among molluscs, 77 freshwater species and 270 species of land snails are from this region. Nearly 10 per cent of total area (13,595 sq. km) of Western Ghats is currently covered under the Protected Area category which includes 52 sanctuaries, 15 National parks, 9 Tiger Reserves and 2 Biosphere reserves (Emiliyamma and Palot, 2012). The Western ghats has an impressive diversity of reptiles including one of the most dangerous snake of Western Ghats, a Viper. There are 13 species of lizards found in Western Ghats, few of them are monitor lizard, dwarf gecko and the malabar flying lizard. The great Western Ghats is also home for the largest venomous snake of world and biggest venomous snake in India, King Cobra. The giant King Cobra inhabits and prays into the endangered rain forests of the Western Ghats to the dense forest of Kerala. Western Ghats is one of the global biodiversity hotspots which have the wonderful world of insects. There are around 6000 insect species found in the Western Ghats and Nilgiri biosphere reserve area. Due to the significance of biodiversity richness, UNESCO has included the Western Ghats as a 'World Heritage Site' in July 2012.

Faunal Conservation Strategies

The countries adopted an ambitious Strategic Plan at the Convention of Biological Diversity (CBD) 2011-2020 in October, 2010 at Nagoya, Japan with a 20 time bound target to halt the extinction rate by 2020. These target popularly known as 'Aichi Targets' clearly defines that by 2020, the extinction of threatened species should be prevented and their conservation status of those most decline has been improved and sustained. According to UN FAO prediction, 24 per cent of world's mammalian population will be extinct by 2020. Nearly 12 per cent of Indian bird population also faced the extinct condition. The Bird Life International and IUCN have listed 12 species of birds as 'critically endangered' category. The conservation strategies and the utilization of biodiversity for sustainable development of nation includes special protection to biodiversity by establishment of national parks, wildlife sanctuaries, biosphere reserves, ecologically fragile and sensitive areas. Others strategies include afforestration of degraded areas and wastelands and establishment of gene banks as part of *ex-situ* conservations. The four forest types (tropical, sub-tropical, temperate and alpine) were further divided into sixteen major types and 232 sub-types for the purpose of conservation. The major *in-situ* conservation of India lies in its impressive Protected Areas (PAs) network. A network of 668 PAs has been established, extending over 1, 61, 221.57 sq. kms (4.90 per cent of total geographic area), comprising 102 National Parks, 515 Wildlife Sanctuaries, 47 Conservation Reserves and 4 Community Reserves (MoEF, 2011a). This network resulted in significant restoration of large population of mammals such as tiger, lion, rhinoceros, crocodiles, and elephants. In addition, for special flagship programmes of specific management of tiger and elephant habitats, overall 40 Tiger Reserves and 28 Elephant Reserves have been established. The Ministry of Environment and Forests (MoEF) established the National Afforestation and Eco-development Board (NAEB) in August 1992 to promote afforestation and eco-development strategies for join forest management and microplanning. The various central Acts related to biodiversity includes Forest Act, 1927, Wildlife (Protection) Act, 1972, Forest (Conservation) Act, 1980 and the Environment (Protection) Act, 1986 and the policies and strategies directly relevant to biodiversity includes National Forest Policy (1988), National Conservation Strategy and Policy Statement for Environment and Sustainable Development, National Agricultural Policy, National Land

Use Policy, National Fisheries Policy, National Policy and Action Strategy on Biodiversity, National Wildlife Action Plan and Environmental Action Plan.

References

1. Andrews, H. (2006) Status of the Indian Gharial (*Gavialis gangeticus*), Conservation Action and Assessment of Key Locations in North India. Unpublished Report to the Madras Crocodile Bank Trust.

2. Archibald, G.W. and Meine, C.D. (1996) Family Gruidae (Cranes). *In*: Auks, H., del Hoyo, J., Elliott, A. and Sargatal, J. (Eds), Handbook of the Birds of the World, Volume 3. Lynx Edicions, Barcelona, Spain.

3. BirdLife International (2010) *Houbaropsis bengalensis*. *In*: IUCN 2011. IUCN Red List of Threatened Species. Version 2011.2. www.iucnredlist.org.

4. Borang, A. (2004) Birds of Arunachal Pradesh in northeast India – the checklist and distribution in protected areas. Forest Zoology Division, SFRI, Itanagar.

5. Chatterjee, S., Saikia, A., Dutta, P., Ghosh, D., Pangging, G. and Goswami, A.K. (2006) Biodiversity significance of northeast India. WWF-India, New Delhi.

6. Choudhury, A. (2002) The Red Panda – Status and Conservation. *In*: Chatterjee, S. (Eds), Biodiversity 'Hot spots' Conservation Programme (BHCP) Final Report Vol II, WWF-India, New Delhi. pp 132-168.

7. Critically Endangered Animal Species of India, March 2011, Ministry of Environment and Forests and Zoological Survey of India.

8. Dandapat, A., Banerjee, D. and Chakraborty, D. (2010) The case of the disappearing house sparrow (*Passer domesticus indicus*). Vet. World 3: 97-100.

9. Datta, A., Pansa, J., Madhusudan, M.D. and Mishra, C. (2003) Discovery of the leaf deer *Muntiacus putaoensis* in Arunachal Pradesh: an addition to the large mammals of India. Curr. Sci. 84: 101-103.

10. Deccan Herald (2012) Population of Ganges River Dolphin on the decline, October 3, 2012.

11. Devashish Kar, Nagarathna, A.V. and Ramachandra, T.V. (2000) Fish diversity and conservation aspects in an aquatic ecosystem in India. Conference: INTECOL VI: Wetlands in the millennium, 6-12, August 2000, Quebec, Canada, Organised by International Association of Ecology. (http://www.sws.org/quebec2000/abstracts_ program/index.html).

12. Emiliyamma, K.G. and Palot, M.J. (2012) Fauna of protected areas of Western Ghats. Zoological Survey of India, Kolkata.

13. Faizuddin, M. (1985) Distribution, abundance and conservation of Gharials in Bangladesh. Tiger Pap., 12: 22-23.

14. Global Tiger Initiative (2011) Global Tiger Recovery Program 2010-2022. Global Tiger Initiative Secretariat, The World Bank, Washington DC, USA.

15. Gole, P. (1989) The status and ecological requirements of Sarus Crane - Phase I. Ecological Society, Pune, India.

16. Goyal, A.K. and Arora, S. (2009) India's Fourth National Report to the Convention on Biological Diversity. Ministry of Environment and Forests, Government of India, New Delhi.

17. Hussain, A. (2012) Kashmir to raise endangered Hangul in captive breeding center. Hindustan Times, September 16, 2012.

18. Indian Rhino Vision 2020, Department of Environment and Forests, Government of Assam.

19. IUCN (2008) IUCN Red List of Threatened Species. IUCN Species Survival Commission. Gland, Switzerland (www.iucnredlist.org).

20. Jaiswal, D.P. and Ahirrao, K.D. (2012) Icthyodiversity of the Rangavali Dam, Navapur District, Maharashtra State. J. Res. Biol. 3: 241-245.

21. Jones, S. and Kumaran, M. (1980) Fishes of the *Laccadive archipelago*. (Kerala The Nature Conservation and Aquatic Sciences Service, Trivandrum), pp. 760.

22. Kaul, R. (2000) Arunachal. In quest of nature's bounty. WWF-India, New Delhi.

23. Khan, M.A.R. (1979) Gharial extinct in Bangladesh. Crocodile Specialist Group Newsletter 1: 2.

24. Lal, K.K. and Jena, J.K. (2011) Vision 2030, National Bureau of Fish Genetic Resources, ICAR, Lucknow.

25. Laurance, W.F. (2007) Forest destruction in tropical Asia. Curr. Sci. 93: 1544-1550.

26. Maskey, T.M. (1999) Status and Conservation of Gharial in Nepal, *In: ENVIS (Wildlife and Protected Areas)* I, Vol. 2(1). Wildlife Institute of India, Dehra Dun, pp. 95–99.

27. McCarthy, T.M. and Chapron, G. (2003) Snow Leopard Survival Strategy. International Snow Leopard Trust and Snow Leopard Network, Seattle, USA.

28. Ministry of Environment and Forests (2006) Action Plan for Vulture Conservation in India, April 2006, Government of India, New Delhi.

29. Ministry of Environment and Forests (2011a) Report to the People on Environment and Forests 2010-2011, Government of India, New Delhi.

30. Ministry of Environment and Forests (2011b) India Tiger Estimate 2010, Government of India, New Delhi.

31. Ministry of Environment and Forests (2012) Annual Report 2012, Government of India, New Delhi.

32. National Biodiversity Action Plan (NBAP) (2008) Ministry of Environment and Forests, Government of India, New Delhi.

33. Rangarajan, M., Desai, A., Sukumar, R., Easa, P.S., Vivek Menon, Vincent, S., Suparna Ganguly, Talukdar, B.K., Brijendra Singh, Divya Mudappa, Sushant Chowdhary and Prasad, A.N. (2010) Gajah – Securing the future for elephants in India. The Report of the Elephant Task Force, Ministry of Environment and Forests, Government of India, New Delhi.

34. Saikia, B.P., Saud, B.J., Kakati Saikia, M. and Saikia, P.K. (2010) Present distribution status and conservation threats of Indian Gharial in Assam, India. Int. J. Biodiversity Conser. 2: 382-387.

35. Sankaran, R. (1993) Red data bird: Lesser Florican. World Birdwatch 15: 18-19.

36. Sankaran, R. (1997) Habitat use by the Lesser Florican in a mosaic of grassland and cropland: the influence of grazing and rainfall. J. Bombay Nat. His. Soc. 94: 40-47.

37. Singh, L.A.K. (1991) Distribution of *Gavialis gangeticus*. Hamadryad 16:39–46.

38. Sinha, R.K., Dubey, M., Tripathi, R.D., Kumar, A., Tripathi, P. and Dwivedi, S. (2010). India as a megadiversity nation. Environews 16: 1-3.

39. Strassburg, B.B.N., Rodrigues, A.S.L., Gusti, M., Balmford, B., Fritz, S., Obersteiner, M., Turner, R.K. and Brooks, T.M. (2012) Impacts of incentives to reduce emissions from deforestation on global species extinctions. Nature Climate Change.

40. Thompson, C. (2009) The Eastern Himalayas: Where worlds collide. WWF-India.

41. Tyagi, L.K. and Jena, J.K. (2012) Annual Report 2011-2012, National Bureau of Fish Genetic Resources (ICAR), Lucknow.

42. Venkataraman, K and Wafar, M. (2005) Coastal and marine biodiversity of India. Ind. J. Mar. Sci. 34:57-75.

43. Virani, M. (2006) Asian Vulture Crisis Project: Pakistan, International Conference on Vulture Conservation, New Delhi, 31 Jan-1 Feb, 2006.

44. Whitaker, R. (1976) Gharial survey report. Mimeographed report for the New York, Zool. Soc., pp.1-19.

45. Whitaker, R. (2007) The Gharial: Going extinct again. Iguana 14:25-33.

46. Whitaker, R. and Basu, D. (1983) The Gharial (*Gavialis gangeticus*): A review. J. Bombay Nat. His. Soc. 79:531–548.

47. WWF-India (2012) Protecting Bustards in India. Special Issue, September 2012.

2013, Glimpses of Animal Biodiversity   *Pages* 19–28
Editor: **Dr. K. Muthuchelian,** *Vice Chancellor, Periyar University, Salem*
Published by: Daya Publishing House, NEW DELHI

Chapter 2

Phytoplanktonic Composition, Abundance and Diversity in the Semi-Arid Coastal Waters of Mundra, Kachchh, Gujarat

V. Devi and G.A. Thivakaran*

Gujarat Institute of Desert Ecology,
P.B. No. 83, Mundra Road, Bhuj – 370 001, Kachchh, Gujarat

ABSTRACT

Tide wise phytoplanktonic density, diversity, composition and seasonal shift in the near coastal waters of Mundra, a semi-arid coastal belt at the Kachchh district of Gujarat were investigated along with 7 crucial water quality parameters. This study was conducted at four sites in a radius of 10 km in close vicinity of Adani port complex and a special economic zone (SEZ). In total, 99 species in 46 genera were recorded during this winter and pre and post-monsoon seasonal investigation. Seasonal shift in composition was moderate with post-monsoon recording the highest taxa of 70 species followed by winter (67 species) and pre monsoon (63 species). Diatoms were the dominant group in the phytoplankton community, contributing a maximum of 77 per cent followed by dinoflagellates (6 per cent), Haptophyceae (1 per cent) and 16 per cent other minor groups in the overall annual composition. Overall seasonal density values were comparatively less and ranged from a minimum of 98 cells/l and maximum of 368 cells/l during winter, pre-monsoon and post-monsoon, respectively at all the four stations. Density was marginally higher during low tide than high tide

* Corresponding Author: E-mail: devi_marine08@rediffmail.com

probably due to nutrient enrichment and presence of mangrove ecosystem in the vicinity. Shannon-weiner diversity indices values were higher during winter and ranged from 2.67 to 3.48 while during other seasons the values were marginally less. Variance of phytoplankton density and diversity with environmental parameters were tested with 2 Way ANOVA which showed that majority of parameters had a significant negative correlation at 0.05 levels. Results were discussed in the light of ongoing industrial activity at Mundra.

Keywords: *Mundra, Phytoplankton, Density, Diversity, Composition.*

Introduction

Phytoplankton is the primary producer of the ocean from which the energy is synthesized and transferred to higher organisms through food chain. Phytoplankton production contributes nearly 95 per cent of the total organic production in marine ecosystem. The rate of gross primary productivity determines fisheries yield especially in the coastal waters. Hence estimation of abundance of phytoplankton and its season shift in composition is the best means for quantitative assessment of potential fisheries of an area. (Sithik *et al.,* 2009). They also serve as bio-indicators with reference to water quality and thus act as a tool for assessing the health of the aquatic ecosystems. (Rajkumar *et al.,* 2009).

Study on the phytoplanktonic distribution and composition in Kachchh region is very limited (Nair 2002; Saravanakumar *et al.,* 2008). However, information on the phytoplankton community structure and diversity in the waters of west coast of India are many (Achuthankutty *et al.* 1981; Jayalakshmy *et al.,* 1986; Davassy and Goes, 1989; Ramaiah *et al.,* 1995; Nair, 2002; Madondkar *et al.,* 2007). Kachchh coastal waters present several physical, chemical and biological uniqueness due to its poor rainfall and high aridity (3 in a scale ranging from 1 to 4). Near absence of any scientific studies combined with its ecological uniqueness prompted us to undertake the present seasonal study on the composition, abundance and distribution of phytoplankton and their tide-wise variation.

Study Area

Coastal stretch of Kachchh district extends for about 405 km constituting the whole northern coast of Gulf of Kachchh (GoK) (Figure 2.1). Mudflats (940 km^2), mangroves (727 km^2), sandy shores and creek systems characterize Kachchh coast. Due to its semi-arid nature, annual rainfall in Kachchh coast is poor, ranging from 250-350 mm which is often irregular. Mundra taluka, where the study sites are located normally receives an average rainfall of 407 mm (1932 to 2001). Rain during monsoon is confined to only 15-20 days and occurs as an instant downpour. Freshwater input into the near coastal waters is quite meager and appears to have least influence on the ambient coastal water quality except during monsoon months. Annual temperature ranges from 7–48°C with a yearly average humidity of 60 per cent. Tides in Kachchh are mixed and predominantly semidiurnal with Mean Highwater Spring (MHWS) of 6.66 m and Mean Highwater Neap (MHWN) of 5.17 m. Coastal stretch of Mundra is marked by narrow beaches and wide mudflats with predominantly muddy alluvial substrate. Mangrove extent in the vicinity of the study area is about 19.1 km^2 distributed mostly along the creek systems. Of late, coastal zone of Mundra is undergoing aggressive industrial development like port, thermal power plants and India's largest coastal Special Economic Zone (SEZ).

Sampling Method

Surface water and phytoplankton samples were collected in the port area at four sites [Bocha Creek (BC), Jetty East (JE), Jetty West (JW) and Baradimatha creek (BMC) (Map 2.1)] in a radius of around 10 km during winter, pre-monsoon and post monsoon seasons of 2005. Surface water temperature was measured *in-situ* with Merck mercury thermometer of 0.1°C accuracy and pH was measured with a hand held, buffer calibrated pH meter. Salinity was estimated by refractometer (Atago, Japan) and Dissolved oxygen was estimated by Winklers titrimetric method (Strickland and Parsons, 1972). Nutrients in surface water samples were analysed as described by Strickland and Parsons (1972). Total Suspended Solids (TSS) was estimated by filtering one litre of water in a Whatman filter paper (No.42) and the results were expressed as mg/l. Turbidity measurements were made with a field based Hanna Nephelometer (HI-93703).

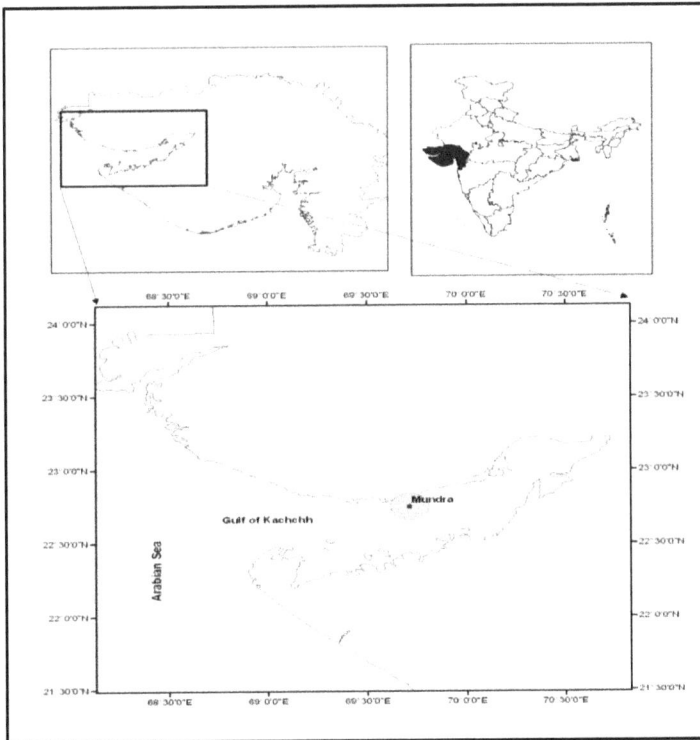

Map 2.1: Sampling location in Kachchh coast.

Phytoplankton samples were collected using standard plankton net with a mesh size of 51μm and a mouth area of 0.1256 m² (20 cm radius). The net fitted with a flow meter (Hydrobios) was towed from a motorized boat. Plankton collected from the net was preserved with 5 per cent neutralized formaldehyde for further analysis. The initial and final flowmeter reading was noted down for calculating the amount of water filtered. Quantitative and qualitative analysis of phytoplankton (cell count) was carried out using Sedgewick-Rafter counting chamber. Density (no/l) was calculated using the following formula:

$$N = n \times v/V$$

where,

N is the total no/l; n is average no of cells in 1 ml; v is volume of concentrate; V is total volume of water filtered.

Species diversity index (H') was calculated using Shannon and Weiner's (1949). Species richness and evenness was calculated using Simpons index and Pielou's Evenness (Gleason, 1992; Pielou, 1996). Two Way Analysis of Variance (ANOVA) was performed to test the homogeneity of various physico-chemical parameters and the relationship among them. Simple correlation co-efficient analysis was also employed for further statistical interpretation of data.

Results

Results of seven major water quality parameters analyzed are given in Table 2.1. Surface water was hypersaline in all the three seasons and varied from 37 to 41 ppt. Temperature was highest (30°C) during pre-monsoon (June) and lowest (23°C) during winter (January). In all seasons pH value ranged between 7.9 and 8.3, showing alkaline nature of the surface water. Recorded dissolved oxygen levels were fairly good with a range of 3.397 mg/l to 6.014 mg/l during pre-monsoon month of June and winter month of January. Minimum turbidity value of 11.06 NTU was recorded during January and the maximum value (228 NTU) was during June. Maximum total suspended solid level of 256 mg/l was recorded during pre-monsoon and minimum level was in winter (6 mg/l). Lowest level of nitrate content was observed in winter (11 μg/l) and the highest level (155 μg/l) was during post-monsoon. Nitrite values ranged from 6 μg/l during pre-monsoon to 40 μg/l during winter. Phosphate levels fluctuated between 15μg/l to 442 μg/l recorded respectively during pre-monsoon and winter. Except salinity and dissolved oxygen, tide-wise pattern could not be discerned in the studied parameters.

Analysis of Variance results were not significant between seasons and stations for salinity, temperature and nitrite while significant relationship was shown between other parameters like dissolved oxygen, TSS, nitrate and phosphate.

Season wise variation of phytoplankton genera and species distribution for the four sites is shown in Figure 2.1. In total, 99 species in 46 genera were recorded across seasons, stations and tides. Seasonal shift in composition was moderate with post-monsoon recording the highest taxa of 70 species followed by winter (67 species) and pre-monsoon (63 species). Diatoms were the dominant group in the phytoplankton community, contributing a maximum of 77 per cent followed by

Table 2.1: Seasonal (2005) surface water quality of the study sites.

Seasons	Salinity–ppt							
	BC		JE		JW		BMC	
	Lt	Ht	Lt	Ht	Lt	Ht	Lt	Ht
Winter	39.5	39	39	38.5	39	38.5	40	39.5
Pre-Monsoon	41	40	40	39.5	40	39	41	40
Post-Monsoon	37	38	38	38	39	38	38	37
Temperature (°C)								
Winter	23	23.5	23	24	23	24	23	24.5
Pre-Monsoon	29	29	29	29	29	28	30	29
Post-Monson	27	26	28	27	28	28	26	26
pH								
Winter	8.3	8.2	8.2	8.2	8.3	8.2	8.2	8.2
Pre-Monsoon	8.1	7.9	8	8.1	8	8	8	8.1
Post-Monsoon	8	8.2	8	8.1	8.1	8.1	8.1	8.1
DO (mg/l)								
Winter	5.621	6.014	5.556	5.752	5.752	5.883	5.425	5.818
Pre-Monsoon	3.545	3.444	3.397	3.697	4.102	4.26	3.697	3.854
Post-Monsoon	3.85	4.27	3.55	3.95	4.25	4.56	3.92	4.85
Turbidity (NTU)								
Winter	27.67	66	11.06	99	18.77	75	84	49
Pre-Monsoon	143	205	103	227	166	228	171	167
Post-Monsoon	96	126	94	128	136	175	125	156
TSS (ppm)								
Winter	14	32	28	76	6	76	52	34
Pre-Monsoon	118	103	256	175	120	171	149	144
Post-Monsoon	74	87	65	69	36	54	45	62
Nitrate–Nitrogen (µg/l)								
Winter	31	41	0	0	0	11	43	15
Pre-Monsoon	140	70	120	56	40	80	150	70
Post-Monsoon	89	65	45	37	58	64	25	155
Nitrite–Nitrogen (µg/l)								
Winter	137.33	181.63	0	0	0	48.73	190.49	66.45
Pre-Monsoon	18	11	14	14	7	10	15	6
Post-Monsoon	21	10	16	15	9	12	13	11
Phosphate–Phosphorus (µg/l)								
Winter	286	394	236	314	312	442	342	313
Pre-Monsoon	31	15	21	25	27	28	21	34
Post-Monsoon	43	31	36	28	38	43	26	34

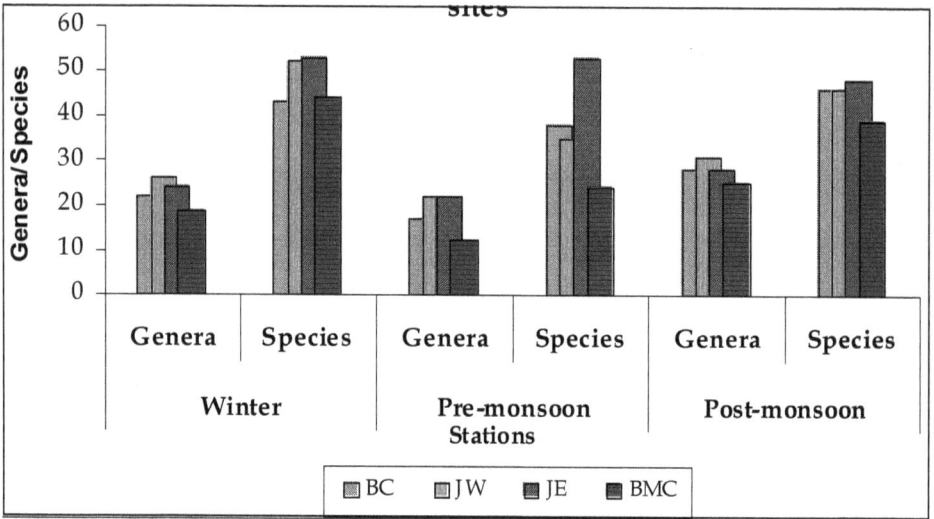

Figure 2.1: Distribution of phytoplankton in the study sites.

dinoflagellates (6 per cent), Haptophyceae (1 per cent) and 16 per cent other minor groups in the overall composition. Among various groups, *Biddulphia, Cheatoceros, Coscinodiscus, Navicula, Nitzchia, Pleurosigma,* and *Rhizosolenia* were most dominant. Phytoplankton showed that density at BM was highest during all the seasons. However, the minimum and maximum densities at each station varied considerably during different seasons. Overall seasonal density values were comparatively less and ranged from a minimum of 98 cells/l and maximum of 368 cells/l during winter, pre monsoon and post-monsoon, respectively at all the four stations. Tide specific variation in phytoplankton density values were obvious with low tide always recording higher phytoplankton density values than high tide at most of the stations, except at BC during winter and at JW during pre-monsoon (Table 2.2).

Species Diversity, Evenness and Richness

Shannon-weiner diversity indices values for phytoplankton ranged from 2.26 to 3.48 bits/individual (Table 2.3). Species evenness index values ranged from 0.68 to 0.96 and species richness indices varied from 4.4 to 8.63 during winter, pre monsoon and post-monsoon, respectively. The statistical analysis did not show any discerned pattern during the low and high tide. Observed values indicated the healthy conditions and natural seasonal fluctuations with regard to phytoplankton community in the study area.

Discussion

Planktonic communities are influenced by physical factors like tidal cycle, continuous changes in salinity and temperature and other seasonal physico-chemical variations. This leads to temporal and spatial variation in number of species, associated with an increase in the abundances of those few species which are well adapted to changing environmental parameters (Odum and Odum, 1959). Coastal waters of arid

Table 2.2: Phytoplankton density and no. of genus/species at study site.

Season	Stn	Tide	No of Genus/ Species	Density	Station-wise Density	Seasonal Density	Low Tide Density	High Tide Density
Winter	BC	LT	31/18	201	253	263.4	323.7	203
		HT	38/19	305				
	JW	LT	49/21	360	249.5			
		HT	31/16	139				
		LT	40/20	366	259.5			
	J E	HT	42/20	153				
	BM	LT	33/14	368	291.5			
		HT	36/17	215				
Pre-Monsoon	BC	LT	17/37	253	246	196.2	212.2	180.2
		HT	17/38	239				
	JW	LT	18/27	98	107.5			
		HT	22/31	117				
	J E	LT	22/46	175	166			
		HT	16/32	157				
	BM	LT	13/22	323	265.5			
		HT	10/17	208				
Post-Monsoon	BC	LT	34/24	212	203	189.25	200.7	177.8
		HT	35/28	194				
	JW	LT	33/26	127	117.5			
		HT	31/25	108				
	J E	LT	32/26	152	140.5			
		HT	34/25	129				
	BM	LT	24/19	312	296			
		HT	29/20	280				

zones like Kachchh are likely to respond to environmental changes in different ways than the other coastal systems receiving normal rainfall. In Kachchh coast, low precipitation, high salinity and flash floods during short spell of terrestrial run-off expose them to sudden shift in governing parameters and render them highly vulnerable, favouring only hardy species capable of withstanding harsh environmental conditions. Dominance of few genera in the present study attest this fact Station-wise composition did not register significant difference and the dominant species recorded were highly tolerant of the prevailing ambient conditions. Low density values (98 to 368 cells/l) for the entire study period may be due to these fast shifting environmental factors. Recorded density values seems to be very less considering the phytoplankton densities reported earlier along west coast waters (Tiwari and Nair, 1998; Ramiah *et al.*, 1998; Selvaraj *et al.*, 2003). Fluctuations in phytoplankton population densities were due to various

Table 2.3: Diversity, richness and evenness indices for phytoplankton in the study area.

Season	Stations	Tides	Diversity-H'	Evenness	Richness
Winter	BC	LT	2.67	0.78	5.86
	BC	HT	3.05	0.83	6.88
	JW	LT	2.95	0.76	8.24
	JW	HT	2.93	0.86	5.87
	JE	LT	2.8	0.8	6.7
	JE	HT	3.5	0.9	8
	BM	LT	3.3	0.95	7.74
	BM	HT	3.33	0.93	7.77
Pre-Monsoon	BC	LT	3.32	0.68	6.52
	BC	HT	3.24	0.93	6.48
	JW	LT	2.94	0.89	5.81
	JW	HT	2.55	0.75	5.37
	JE	LT	3.39	0.89	8.63
	JE	HT	3.03	0.88	6.39
	BM	LT	2.26	0.73	4.45
	BM	HT	2.49	0.88	4.11
Post-Monsoon	BC	LT	3.36	0.95	5.89
	BC	HT	3.23	0.91	5.99
	JW	LT	3.32	0.95	5.74
	JW	HT	3.28	0.96	5.67
	JE	LT	3.17	0.92	6.2
	JE	HT	3.26	0.92	6.42
	BM	LT	3.05	0.96	4.4
	BM	HT	3.1	0.9	5

LT: Low tide; HT: High tide.

factors such as salinity, light, turbidity, temperature and nutrients. (Chandran 1987; Roden *et al.,* 1987). Considering the given climatic conditions at the study site, it is reasonable to have lesser phytoplankton density. Low density observed during monsoon in the present study could be due to increased total suspended solids (TSS) brought in by terrestrial run-off. This increased TSS level could also be due to the combination of high wave actions during winter and the dredging activities of the nearby port. While tide-wise, generic and species composition were almost similar, density recorded a clear variation with low tide recording marginally higher values than high tide which could be due to the presence of mangrove ecosystem in the vicinity.

References

Achuthankutty, C.T., Nair, S.R.S., Devassy, V.P. and Nair, V.R., 1981. Plankton composition in two estuaries of Konkan coast during pre-monsoon season. *Mahasagar*, 14(1): 55–60.

Chandran, R., 1987. Hydrological studies in the gradient zone of the vellar estuary Benthic fauna. *Mahasagar*, 24(2): 81–88.

Davassy, V.P. and Goes, J.I., 1989. Seasonal patterns of phytoplankton biomass and productivity in a typical estuarine complex. *Proc. Indian Acad. Sci.*, 99(55): 485–501.

Gleason, H.A., 1922. On the relation between species and area. *Ecology*, 3: 156–162.

Jayalakshmy, K.V., Kumaran, S. and Vijayan, M, 1986. Phytoplankton distribution in Cochin backwaters: A seasonal study. *Mahasagar*, 19: 29–37. NIO, Goa.

Krishna Kumari, L. and John, Julie, 2003. Biomass and quantitative indices of phytoplankton in Mandovi–Zuari estuary. NIO, Goa.

Madondkar, S.G.P., Gomes, H., Parab, S.G., Pednekar, S. and Goes, J.I., 2007. Phytoplankton diversity, biomass, and production. In: *The Mandovi and Zuari Estuaries*, (Eds.) S.R. Shetye, M. Dileep Kumar and D. Shankar, pp. 67–81.

Mathivanan, V., Vijayan, P., S. Sabhanayakam and Jeyachitra, O., 2007. An assessment of plankton population of Cauvery river with reference to pollution. *J. Environ. Biol.*, 28: 523–526.

Sithik, M.A., Thirumaran, G., Arumugam, R., Ragupathi Raja Kannan, R. and Anantharaman, P., 2009. Studies of phytoplankton diversity from Agnitheertham and Kothandaramar Koil coastal waters, Southeast coast of India. *J. Environ. Res.*, 3(2): 18–125.

Nair, V., 2002. *Status of Fauna and Flora in Gulf of Kachchh*. National Institute of Oceanography, Goa, pp. 144.

Odum, H.T. and Odum, E.P., 1959. *Fundamentals of Ecology*. W.B. Saunders, Philadelphia, PA, pp. 536.

Pielou, E.C., 1966. The measurement of diversity in different types of biological collection. *J. Theoret. Biol.*, 13: 144.

Rajkumar, M., Perumal, P., Prabhu, V. Ashok, Perumal, Vengadesh and Rajasekar, K. Thillai, 2009. Plankton diversity in Pichavarm mangrove waters from south-east coast of India. *Environ. Biol.*, 30(4): 488–498.

Ramaiah, N., Ramaiah, N., Chandramohan, D. and Nair, V.R., 1995. Autotrophic and heterotrophic characters in polluted esturine complex. *Estn. Coast Shelf Sci.*, 40(1): 45–55.

Ramiah, N. Ramiah and Nair, V.R., 1998. Phytoplankton characteristics in a polluted Bombay harbor–Thana–Bassein creek estuarine complex. *Indian J. Mar.Sci.*, 27: 281–285.

Roden, C., Roadhouse, M.P.G and Hendry, M.P., 1987. Hydrography and the distribution of phytoplankton in Killar harbor: A Fjord in Western Ireland. *J. Mar. Biol. Ass.*, U.K., 67: 359–371.

Saravanakumar, A., Rajkumar, M., Thivakaran, G.A. and Serebiah, J. Sesh, 2008. Abundance and seasonal variations of phytoplankton in the creek waters of western mangrove of Kachchh, Gujarat. *J. Environ. Biol.*, 29: 271–274.

Selvaraj, G.S.D., Thomas, V.J. and Khambadkar, L.R., 2003. Seasonal variation of phytoplankton productivity in the surf zone and back water at Cochin. *J. Mar. Biol. Assoc., India*, 45(1): 9–19.

Shannon, C.E. and Weaver, W., 1949. *The Mathematical Theory of Communication*. University of Ilinois Press, Urbana, pp. 117.

Shashi Shekhar, T.R., Kiran, B.R., Puttaiah, E.T., Shivaraj, Y. and Mahadevan, K.M., 2008. Phytoplankton as index of water quality with reference to industrial pollution. *J. Environ. Biol.*, 29: 233–236.

Strickland, J.D.H. and Parsons, T.R., 1972. A practical handbook of seawater analysis. *Bull. Fish. Res. Bd., Canada,* 167: 311.

Tas, Beyhan and Gonulol, Arif, 2007. An ecologic and taxonomic study on phytoplankton of a shallow lake, Turkey. *J. Environ. Biol.*, 28: 439–445.

Tiwari, L.R. and Nair, Vijayalakshmi R., 1998. Ecology of phytoplankton from Dharmatar Creek, West coast of India. *Indian J. Mar. Sci.*, 27: 302–309.

Tiwari, A. and Chauhan, S.V.S., 2006. Seasonal phytoplanktonic diversity of Kitham lake, Agra. *J. Environ. Biol.*, 27: 35–38.

2013, Glimpses of Animal Biodiversity

Editor: **Dr. K. Muthuchelian,** *Vice Chancellor, Periyar University, Salem*

Published by: Daya Publishing House, NEW DELHI

Pages **29–39**

Chapter 3

Phytoplankton Diversity of Yelebethur Pond Near Davangere City, Karnataka State: A Seasonal Study

Nafeesa Begum[1] * and J. Narayana[2]

[1]*Department of Botany, Sahyadri Science College (Auto), Shimoga*
[2]*Department of Environmental Science,*
Kuvempu University, Shankaraghatta

ABSTRACT

A study was carried out in Yelebethur pond near Davangere city, Karnataka on phytoplankton diversity, density and distribution in different seasons and their correlations with physico-chemical properties of water. A total of 56 phytoplankton taxa belonging to Chlorococcales, Blue-greens, Desmids, Diatoms and Euglenoids were represented. Relative abundance of phytoplankton in Yelebethur pond showed maximum of Chlorococcales (46.65 per cent), followed by Blue-greens (43.30 per cent) followed by Diatoms (9.60 per cent), Euglenoides (0.27 per cent) and Desmids (0.18 per cent). The highest density of phytoplankton was recorded during summer season. Chlorococcales varied with peak density (16,267 org/l) during summer and lowest in rainy season (11,778 org/l), Blue-greens recorded 16,351 org/l and least during winter with 14,289 org/l. Diatoms varied (4,622 org/l) during summer and minimum with (1,822 org/l) during rainy season, Desmids varied from 92 org/l during rainy season and lowest during rainy season with 28 org/l, Euglenoids recorded 132

* Corresponding Author: E-mail: nafeeza_khaliqh @ yahoo.co.in org/l during summer and least in winter with 67 org/l. Our study revealed that the growth of phytoplankton is governed by BOD, Carbon dioxide, COD, Sulphate, Total alkalinity and Air temperature. Among these parameters of Yelebethur pond, COD is positively correlated with Chlorococcales. Carbon dioxide and Potassium is positively correlated with Blue-green algae. Sulphate is negatively correlated with Desmids. Total alkalinity is positively correlated with Desmids. In this air temperature, sodium, total hardness, chloride and BOD show no correlations. Presence of high density of Chlorococcalean group and Blue-greens indicates eutrophication in the pond. Pollution tolerant species such as *Scenedesmus quadricauda, Coelastrum* sp., *Euglena* sp., *Phacus* sp., *Trachelomonas* sp., *Microcrocystis aeruginosa* are observed.

Keywords*: Yelebethur pond, Phytoplankton diversity, Chlorococcales, Diatoms, Blue-greens, Desmids.*

Introduction

Environmental pollution is a modern day evil affecting all ecosystems including aquatic ecosystems. Therefore the conservation of freshwater environment and its monitoring is highly essential (Mohapatra *et al.,* 1995). Most of the water bodies have become polluted and unfit for human use and the changes in water chemistry are mainly due to mud and silt, rain water dilution, the underlying spring and human activities like washing and bathing (Jeeji Bai *et al.,* 1999). Phytoplankton plays an important role in the biosynthesis of organic matter (primary production) in aquatic systems, which directly or indirectly serve all the living organisms of a water body as food (Anjana S. *et al.,* 1998).

The planktonic study is a very useful tool for the assessment of water quality in any type of water body and also contributes to understanding of the basic nature and general economy of the lake (Pawar *et al.,* 2006). Unplanned urbanization rapid industrialization and indiscriminate use of artificial chemicals in agriculture are causing heavy and varied pollution in aquatic environments leading deterioration of quality and depletion of aquatic biota (Yeole *et al.,* 2005).

Due to certain reasons some planktonic population flair-up to dominate water body and ultimately blooms form. Unlike other algae all the common bloom forming blue-green algae contain gas vacuoles which can impart positive buoyancy to the algae under certain conditions. Some species of blue-green algae aggregate and make a colony floating over the surface forming the bloom. Water bloom besides imparting color to the water also gives a disagreeable smell and taste to it. Phytoplankton species distribution shows wide spatio-temporal variations due to the differential effect of hydrographical factors on individual species and they serve as good indicators of water quality pollution (Gouda *et al.,* 1996).

Phytoplankton of pond ecosystems were studied by Jana (1973), Hosamani and Bharathi (1980), Bhatt and Negi (1985), Saha and Chaudhary (1985), Kant and Raina (1985), Kumar and Dutta (1991), Verma and Mohanty (1995) and several studies on phytoplankton diversity made in India and abroad on the ponds, lakes and reservoirs (Tiwari and Chauhan, 2006; Sridhar *et al.,* 2006; Tas and Gonulol, 2007; Senthikumar and Sivakumar, 2008). In this chapter an attempt has been made to study the seasonal changes and correlation of phytoplankton diversity in relation with physico-chemical parameters in Yelebethur pond.

Materials and Methods

The study was conducted in Yelebethur pond near Davangere city during Sep. 2003 to Aug. 2005. The pond is situated 3 kms away in North East direction of Davangere. It lies between 14° 28' N latitude and 75° 58' E longitudes. Water spread area of this water body is 172 hectares. Rain water is the main source of water and Bhadra right bank channel is the other source. Water is mainly used for irrigation. Cattle bathing and other domestic activities are observed. The catchment area of the water body is 1.11 sq.kms which is covered by areca nut, paddy and natural vegetation.

To evaluate the water characteristics a series of physico-chemical and biological tests were performed during two years period of Sept. 2003 to Aug. 2005. Water samples were collected in Yelebethur pond and mixed as per the standard methods. Sampling at each station consists of taking one litre of sample for biological analysis and two litre in Polyvinyl carbuoys for physico-chemical analysis.

Temperature, pH and DO tests were performed in the field. Alkalinity, chlorides, turbidity and hardness were determined according to standard methods. Phosphate, nitrate, nitrite, sulphate and silica were determined using UV-Visible Spectrophotometer. Standard prescribed methods were followed for the physico-chemical analysis of the water sample (APHA 1998).

For qualitative and quantitative analyses of phytoplankton one liter of composite water samples at surface level were collected at interval of 30 days for 2 years during the period September 2003 to August 2005. One liter of sample was fixed with 20ml of 1 per cent Lugol's Iodine solution and kept 24 hours for sedimentation. 100 ml of sample is subjected to centrifugation at 1500 rpm for 20 minutes and used for further investigation. Identification of plankton up to species level was done by referring standard manuals (Philipose, 1967; Needham and Needham, 1962; Fritch, 1945). Quantitative estimation of phytoplankton was done using by using Sedgewick-Rafter counting cell. The Pearson correlation coefficient was used to examine the relationships among the different environmental variables including phytoplankton density. Correlation coefficient(r) was calculated to detect the relationship between the various parameters of the water bodies under study.

Results and Discussion

The Table 3.1 depicts the Monthly occurrence of different groups of phytoplankton density, Table 3.2 showed, Seasonal variations of phytoplankton density and Table 3.3 showed, Pearsons correlation matrix of different physico-chemical variables. Physico-chemical properties and total phytoplankton density showed significant correlation. Figure 3.1 showed, Seasonal changes of phytoplankton density and Figure 3.2 showed Distribution of phytoplankton percentage (per cent) in Yelebethur pond and Phytoplankton taxa recorded in Table 3.4.

Table 3.1: Monthly occurrence of different groups of phytoplankton density in Yelebethur pond (org/l), September 2003 to August 2005.

Months and Year	Chlorococcales	Diatoms	Desmids	Euglenoids	Blue Greens
Sep-03	14355	2355	38	87	11566
Oct-03	12172	1822	47	78	13155
Nov-03	14089	2178	42	82	11355
Dec-03	12578	2889	58	67	11244
Jan-04	14267	3290	53	78	15688
Feb-04	15644	3467	87	86	13556
Mar-04	14534	4133	92	92	16044
Apr-04	15600	4622	68	88	13867
May-04	16088	3644	72	94	14356
Jun-04	13643	2489	52	72	12711
Jul-04	14045	2044	28	81	13600
Aug-04	14132	2267	32	90	11689
Sep-04	13311	1978	48	83	12667
Oct-04	12267	2133	53	75	11867
Nov-04	13910	1911	46	84	10356
Dec-04	12534	2089	62	71	9689
Jan-05	13956	3378	74	57	11867
Feb-05	14578	3511	81	91	13156
Mar-05	15600	4222	84	106	14356
Apr-05	13956	4533	61	118	12625
May-05	16267	2978	56	132	15867
Jun-05	12756	2400	47	87	13911
Jul-05	11778	2178	32	78	14133
Aug-05	13867	2267	37	88	12756

Table 3.2: Seasonal variations of phytoplankton density in Yelebethur pond (org/l).

Sl.No.	Phytoplankton	Sep. 2003 to Aug. 2004			Sep. 2004 to Aug. 2005			Sep. 2003 to Aug. 2005		
		Winter	Summer	Rainy	Winter	Summer	Rainy	Winter	Summer	Rainy
1.	Chlorococcales	13277	15467	13783	13167	15100	12800	13222	15284	13292
2.	Diatoms	2545	3967	2195	2378	3811	2282	2461	3889	2239
3.	Desmids	50	80	40	59	71	39	55	76	40
4.	Euglenoides	76	90	82	72	112	84	74	101	83
5.	Blue-greens	12861	14456	12667	10945	14001	13600	11903	14229	13134

Table 3.3: Correlation coefficient among the physico-chemical parameters with the density of the phytoplankton in Yelebethur pond.

Parameters	Blue-greens	Chlorococcales	Desmids	Diatoms	Euglenoids
Air temp	0.371	0.272	0.298	0.467	0.482
BOD	0.022	0.200	0.097	0.159	−0.122
Calcium	0.120	−0.346	−0.314	−0.342	−0.234
Chloride	0.251	0.326	0.483	0.427	0.146
COD	0.432	**0.572**	0.233	0.302	0.270
Conductivity	0.276	0.249	0.262	0.220	0.430
Dissolved oxygen	0.097	0.098	0.174	0.264	−0.166
Carbon dioxide	**0.646**	0.255	0.127	0.193	0.414
Magnesium	0.172	0.037	0.148	0.214	0.327
Nitrate	0.252	−0.074	−0.355	−0.090	0.226
Nitrite	0.064	−0.284	−0.495	−0.302	0.030
pH	0.302	0.127	0.184	0.143	0.067
Phosphate	0.169	−0.028	−0.455	−0.057	0.327
Potassium	**0.571**	0.267	0.213	0.370	0.287
Silica	−0.349	−0.237	0.046	−0.205	−0.044
Sodium	0.472	0.093	−0.005	0.219	0.223
Sulphate	−0.047	−0.156	**−0.509**	−0.266	0.061
Total alkalinity	0.257	0.368	**0.633**	0.449	0.077
TDS	0.158	0.242	0.430	0.298	0.300
Total hardness	0.417	0.029	0.143	0.140	−0.086
Turbidity	0.289	−0.066	−0.187	−0.051	0.045
Water temp	0.362	0.206	0.277	0.422	0.402

A total of 56 phytoplankton species were recorded in pond. It constitutes 46.65 per cent of the total phytoplankton population. Chlorococcales represent the first dominated group among the phytoplanktons. The pond comprises of 11 genera and 16 species of Chlorococcales among which *Pediastrum duplex*, *Scenedesmus quadricauda* and *Crucigenia crucifera* were appeared in all months of study period whereas *Chlorella vulgaris* and *Ankistrodesmus* sp. appeared as rare forms.

Species diversity showed that genus *Scenedesmus* was represented by 4 species and *Pediastrum* with 2 species. Similarly the genus *Tetraedron* and *Crucigenia* was represented with 2 species and *Actinastrum*, *Ankistrodesmus*, *Chlorella*, *Chlamydomonas*, *Coelastrum*, *Eudorina* and *Selenastrum* were represented by single species. Some of the pollution tolerant species (Palmer, 1969) identified during the present study are *Scenedesmus quadricauda*, *Coelastrum* sp., *Tetraedon* sp. in water body which indicate eutrophication.

Diatoms in Yelebethur pond represent by 13 genera and 16 species constituting 9.60 per cent of total phytoplankton population. If the diversity of diatoms is considered,

Figure 3.1: Seasonal changes of phytoplankton density of Yelebethur pond and Distribution of phytoplankton percentage (per cent) in Yelebethur pond.

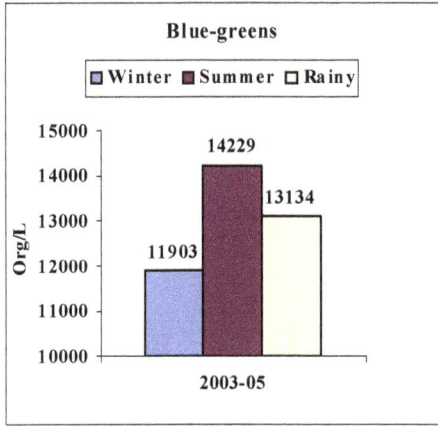

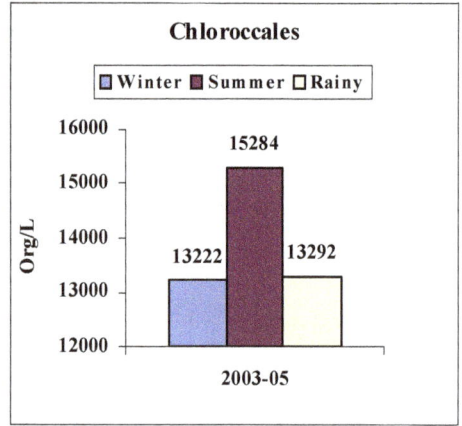

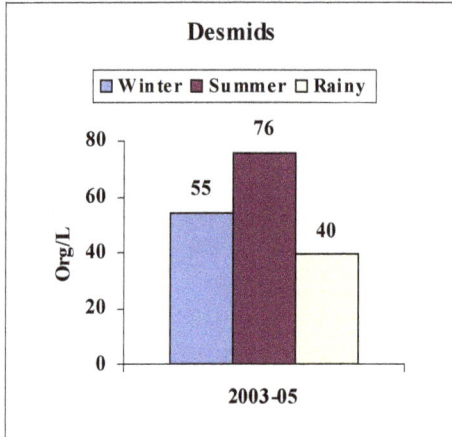

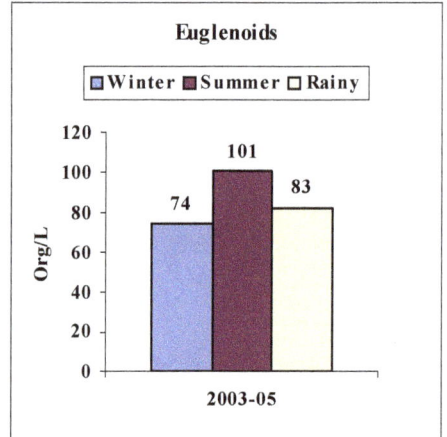

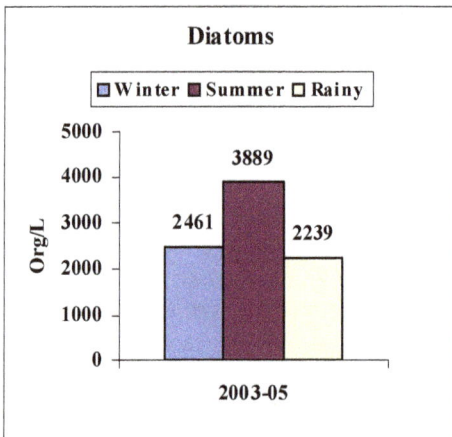

Contd...

Figure 3.1–*Contd...*

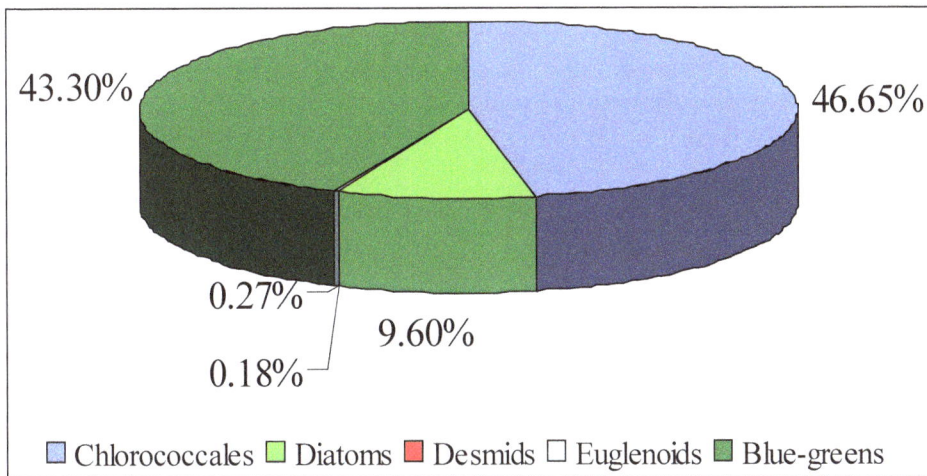

the genus *Navicula* represented by 4 species, *Anamoenies* by 2 species, *Synedra* by 2 species and other genera like *Cymbella*, *Cyclotella*, *Fragillaria*, *Gyrosigma*, *Melosira*, *Surirella*, *Pinnularia* and *Closterium* were represented by single species.

Density of diatoms recorded a maximum of 4622 org/l in the month of April 2004 and a minimum of 1822 org/l during Oct. 2003 whereas seasonal variation of diatoms population showed maximum density during summer season with 3889 org/l and minimum with 2239 org/l during rainy season. Some of the pollution tolerant diatoms (Palmer, 1969) are recorded from the pond are *Synedra* sp., *Cyclotella* sp., *Cymbella* sp. and *Pinnularia* sp.

Desmids represent 5 genera and 7 species constituting 0.18 per cent of total phytoplankton population. When the diversity of desmids is considered Cosmarium was represented by 2 species, *Euastrum* with two species, *Staurastrum*, *Micrasterias* and *Closteriopsis* were represented by single species. The population density reached their peak in the month of March 2004 with 92 org/l, while in the month of July 2004 recorded least number with 28 org/l. Season wise more during summer (76 org/l) followed by winter season (55 org/l) and least in rainy season (40 org/l).

The pond recorded 3 genera and 6 species of Euglenoids and constituting of 0.27 per cent of total phytoplankton population. If the diversity of Euglenoids is considered a genus *Phacus* represented by 3 species, *Euglena* represented by 2 species and *Trachelomonas* represented by a single species. When monthly variations are considered pond recorded a minimum of 67 org/l during the month of December 2003 and a maximum of 132 org/l during May 2005. Seasonal variations showed maximum density with 101 org/l during summer followed by 83 org/l during rainy season and least in winter with 74 org/l. Some of the pollution tolerant species (Palmer, 1969) recorded from the study area are *Euglena* sp., *Phacus* sp. and *Trachelomonas* sp.

Table 3.4: Phytoplankton taxa recorded in Yelebethur pond during Sep 2003-Aug 2005.

Sl.No.	Chlorococcales	Sl.No.	Diatoms
1.	*Actinastrum* sp.	1.	*Anomoeonies sphaeophora*
2.	*Ankistrodesmus falcatus*	2.	*Anomoeonies phybrachysira*
3.	*Arthrodesmus* sp.	3.	*Cymbella cistula*
4.	*Chlorella vulgaris*	4.	*Cyclotella stelligera*
5.	*Coelastrum* sp.	5.	*Fragillaria* sp.
6.	*Crucigenia crucifera*	6.	*Gyrosigma accuminatum*
7.	*Chlamydomonas* sp.	7.	*Melosira granulata*
8.	*Eudorina elegans*	8.	*Navicula pupula*
9.	*Pediastrum simplex*	9.	*Navicula cuspidate*
10.	*Pediastrum duplex*	10.	*Navicula cryptocephala*
11.	*Scenedesmus bijugatus*	11.	*Navicula radiosa*
12.	*Scenedesmus quadricauda*	12.	*Nitzchia amphibia*
13.	*Scenedesmus dimorphous*	13.	*Pinnularia microstauron*
14.	*Scenedesmus bicaudatus*	14.	*Surirella capronii*
15.	*Selenastrum gracile*	15.	*Synedra ulna*
16.	*Tetradaedron longispinum*	16.	*Synedra tabulata*

	Desmids		Euglenoids
1.	*Cosmarium granatum*	1.	*Euglena oxyalis*
2.	*Cosmarium constractum*	2.	*Euglena acus*
3.	*Closteriopsis* sp.	3.	*Phacus longicauda*
4.	*Euastrum sublobatum*	4.	*Phacus circumflexes*
5.	*Euastrum flommeum*	5.	*Phacus pleuronectes*
6.	*Micrasterias* sp.	6.	*Trachelomonas robasta*
7.	*Staurastrum gracile*		

		Blue-greens	
1.	*Agmenellum* sp.	7.	*Nostoc microscopium*
2.	*Anacystis* sp.	8.	*Nostoc microscopium*
3.	*Aphanocapsa* sp.	9.	*Oscillatoria* sp.
4.	*Gloecapsa* sp.	10.	*Phormidium* sp.
5.	*Merismopedia.* sp.	11.	*Spirulina major*
6.	*Microcystis aeruginosa*		

Yelebethur pond supported 11 genera represented by single species of Blue-greens constituting 43.30 per cent of the total plankton population. With regard to their diversity the genus *Anacystis, Anabaenopsis, Aphanocapsa, Agmenellum, Anabaena, Merismopedia, Microcystis, Nostoc, Oscillatoria, Spirulina* and *Gloeocapsa* were represented by single species. Some of the pollution tolerant blue-greens as per

Palmer (1969) were recorded from the study area such as *Oscillatoris* sp., *Microcystis* sp. and *Anabaena* sp., *Microcystis* sp. is used as the best indicator of pollution and associated with highest degree of civic pollution. In the present study *Microcystis aeruginosa* was recorded.

Monthly density of blue-greens recorded minimum of 96589 org/l during December 2004 and maximum of 16,044 org/l during March 2004. Seasonally, summer season recorded more number of blue-greens with 14229 org/l followed during rainy season with 13134 org/l and minimum number during winter season with 11903 org/l.

The data representing correlation coefficient of various parameters (Table 3.3) indicates that among these parameters COD showed positive correlation with chlorococcales. Carbon dioxide showed positive correlation with blue green algae. Potassium was positively correlated with blue green algae. Sulphate showed negative correlation with Desmids. Total alkalinity showed positive correlation with Desmids. In this air temperature, BOD, conductivity and TDS were not showed any correlations with any of the phytoplankton group. Sreenivas and Rana (1994) considered that dissolved oxygen as pre-requisite for the better growth of chlorococcales. However, Hegde and Bharati (1985) concluded that chlorococcales do not show any clear-cut relationship with dissolved oxygen. Hence our results are in confirmation with the findings of Hedge and Bharati.

Manikya Reddy and Venkateswarlu (1992) are of the opinion that dissolved oxygen and blue-greens exhibit a positive correlation with each other. But our studies showed no such correlations. Smith (1983) and Zutchi and Khan (1988) pointed out that high free carbon dioxide favours appreciable number of blue-greens. In our findings Carbon dioxide showed positive correlation with blue green algae.

References

APHA, 1998. *Standard Methods for the Examination of the Water and Wastewater,* 20th Edn. American Public Health Association, Washington, D.C.

Bhatt, S.D. and Negi, Usha, 1985. Physico-chemical features and phytoplankton population in a subtropical pond. *Comp. Physicol. Ecol.*, 10: 85–88.

Fritch, F.E., 1945. *The Structure and Reproduction of Algae*. Cambridge University Press, London, Vol. 2.

Gouda, Rajashree and Panigraphy, R.C., 1996. Ecology of phytoplankton in coastal waters of Gopalpur, east coast of India. *Ind. J. Mar. Science*, 2: 13–18.

Gujarathi, Anjana S. and Kanhere, R.R., 1998. Seasonal dynamics of phytoplankton population in relation to abiotic factors of a fresh water pond at Barwani (M.P.). *Poll. Res.*, 17: 133–136.

Hegde, G.R. and Bharati, 1985. Comparative phytoplankton ecology of fresh water ponds and lakes of Dharwad, Karnataka state, India. *Proc. Nat. Symp. Pure and Appl. Limnology. Bull. Bot. Soc.*, 32: 24–29.

Hosmani, S.P. and Bharati, S.G., 1980. Algae as indicators of organic pollution. *Phykos.*, 1: 23–26.

Jeeji Bai, N. and Lakshmi, D., 1999. Phytoplankton flora of a few temple tanks in Madras and their unique phycobiocoenoses. In: *Land Water Resources–India,* (Eds.) M.K. Durgaprasad and Sankara Pitchaiah. Discovery Publishing House, New Delhi, p. 185–199.

Kant, S. and Raina, A.K., 1985. Linmnological studies of two ponds in Jammu: Qualitative and quantitative distribution of phytoplankton. *Zologica Orintalis,* 2: 89–92.

Kumar, S. and Datta, S.P.S., 1991. Studies on phytoplaktonic population dynamics in Kunjwani pond, Jammu. *Hydrobiologia,* 7: 55–59.

Manikya Reddy, P. and Venkateswarlu, V., 1992. The impact of paper mill effluents on the algal flora of the river Tungabhadra. *J. Indian. Bot. Soc.,* 71: 109–114.

Mohapatra, B.C. and Rengarajan, K., 1995. Effects of some heavy metals copper zinc and lead on certaoin tissues of Liza parsia (Hamilton–Buchanan) in different environments. *CMFRI SPL. Publ.,* 61: 6–12.

Needham, J.G. and Needham, P.R., 1962. *A Guide to Study of Freshwater Biology.* Holden Bay, San Francisco, U.S.A.

Palmer, C.M., 1969. A composite rating of algae tolerating organic pollution. *J. Phycol.,* 5: 76–82.

Pawar, S.K., Pulle, J.S. and Shendge, K.M., 2006. The study on phytoplankton of Pethwaj Dam, Taluka Kandhar, District, Nanded, Maharashtra. *J. Aqua. Biol.,* 21(1): 1–6.

Philipose, M.T., 1967 *Chlorococcales: Monographs on Algae.* ICAR Publications, New Delhi, p. 1–365.

Saha, L.C. and Chaudhary, S.K., 1985. Phytoplankton in relation to abiotic factors of a pond, Bhagalpur. *Comp. Physiol. Ecol.,* 10: 91–100.

Senthikumar, R. and Sivakumar, K., 2008. Studies on phytoplankton diversity in response to abiotic factors in Veeranum lake in the Cuddalore district of Tamil Nadu. *J. Environ. Biol.,* 29: 747–752.

Smith, V.H., 1983. Low nitrogen to phosphorus ratios favour dominance by blue–green algae in lake. *Phytoplankton Science,* 221: 669–671.

Sreenivas, S.S. and Rana, B.C., 1994. Studies on the ecology and tropical status of Gomti village tank in central Gujarat. (India). *Environ. and Applied Biology,* p. 307–314.

Sridhar, R.T., Thangaradjou, T., Senthikumar, S. and Kannan, L., 2006. Water quality and phytoplankton characteristics in the Palk Bay, southeast coast of India. *J. Environ. Biol.,* 27: 561–566.

Tas, B. and Gonulol, A., 2007. An ecologic and taxonomic study on phytoplankton of a shallow lake, Turkey. *J. Environ. Biol.,* 28, 439–445.

Tiwari, A. and Chauhan, S.V.S., 2006. Seasonal phytoplankton diversity of Kitham lake, Agra. *J. Environ. Biol.,* 27: 35–38.

Verma, J.P. and Mohanty, R.C., 1995. Phytoplankton of Malyanta pond of Laxmisagar and its correlation with physico-chemical parameters. *Poll. Res.,* 14: 243–253.

Yeole, S.M. and Patil, G.P., 2005. Physico-chemical status of Yedshi lake in relation to water pollution. *J. Aqua. Biol.,* 20: 41–45.

Zutshi, D.P. and Khan, A.U., 1988. Eutrophic gradient in the Dal lake, Kashmir, *Indian J. Environ. Hlth.,* 30(4): 348–354.

2013, Glimpses of Animal Biodiversity   Pages 40–44
Editor: **Dr. K. Muthuchelian,** *Vice Chancellor, Periyar University, Salem*
Published by: **Daya Publishing House, NEW DELHI**

Chapter 4

Seasonal and Pollution Indicator of Phytoplankton in Lentic Water Bodies of Bhadravathi Taluk, Shimoga District, Karnataka, India

K.V. Ajayan and T. Parameswara Naik

Department of Botany and Environmental Science,
Sahyadri Science College (Autonomous),
Kuvempu University, Shivamogga

ABSTRACT

The study area belongs to Bhadravathi Taluk lies in the central part of Karnataka state, in the south-east corner of the Shimoga district. The latitude and longitude of taluk are 13°N and 75°E with mean elevation from sea level is 608mt. The 288 number of lakes/tanks within 609sq/km areas were recorded and study the phytoplankton diversity with reference to physico-chemical changes. In summer season maximum air temperature 37.5°C and phosphate 2.89mg/l that was favored the dominant species of cyanophyta 3000/l (phytoplankton unit/l) mainly *Merismopedia* spp. But, in case of monsoon season the dominance of species Euglenophyta 1040 org/l; where slightly decreases the physical and chemical parameters that was indicators of seasonal shifting of phytoplankton divisional level. These seasonal indicators in small geographical premises but their occurrence both in seasonal and pollution.

Keywords: *Bhadravathi taluk, Phytoplankton diversity, Physico-chemical parameters.*

Introduction

Phytoplankton

Phytoplankton are generally microscopic they impart green color of the water, phytoplankton per unit area, many of such forms posses other adoptions to aid floating, while turbulence and upward current movements of water keep phytoplankton near surface where photosynthesis is most effective. Phytoplankton community comprises of a heterogeneous group of tiny plants adapted to various aquatic environments. Their nature and distribution varies considerably with respect to seasons and water quality (pollution). Their dominance also leads to quality changes of aquatic systems. Information pertaining to the nature, type and distribution of these organisms provide clue regarding the environmental conditions prevailing in their habitate. According to the size of phytoplankton are classified as Ultra plankton, Nannoplankton, Micro plankton and Macro plankton. The present study revealed that the distribution of phytoplankton mainly on the seasonal as well as the physico-chemical factors (quality water).

Study Area

Geographically, Bhadravathi taluk lies in the central parts of the Karnataka state, in the south-east corner of the Shimoga district. The latitude and longitude coordinates of Bhadravathi town are 13° 50' N and 75° 42'E. Bhadravathi taluk borders, Shimoga taluk of Shimoga district of the west, Honnali taluk of Devangare district of the north, Channagiri taluk of Devangare district of the east and Tarikare taluk of Chikmagalure district to the south. The total area taluk is around 690sq//km and it's at an altitude of around 580m above sea level.

Materials and Methods

For present investigation four sample stations were selected *viz.* BVTS1 which located outs trick of Bhadravathi municipality South West direction and BVTS2 which located within the Bhadravathi municipality area North East direction and other two sample stations such as BVTS3 and BVTS4 lay North West and North East directions. These sample station are directly and indirectly dependence on local community of taluk. The monthly samples collected for phytoplankton analysis was done once in a month from the sample stations for periods of peak summer–May and onset of monsoon–June 2010 for the analysis diversity changes of phytoplankton community and how they related to pollution and season at selected four sample stations. The samples were collected from the surface water by clean glass bottle 30cm away from the water surface; composite sampling was adapted and one liter of composite sample was immediately mixed with 10ml of Lugol's iodine solution and kept for 24 hours for sedimentation. After one day sediment was collected and centrifuged about 15 minutes at 2000rpm. Collected sediment of phytoplankton was made and preserved in 4 per cent formalin solution. The qualitative and quantitative analysis of phytoplankton was done with the help of Lackey's (1938) drop method as mentioned in APHA (1995): This was done shake the concentrated sample of phytoplankton of 1ml and quickly transfer 0.1ml volume of it on a glass slide with the help of a graduated dropper and up to 1ml. Cover the drop carefully with a cover slip. Study a good number of replicates and calculate the average count per 0.1ml.

And using the formula.

$$\text{Phytoplankton (units/L)} = \frac{N \times C \times 10}{V}$$

where,

   N: Number of phytoplankton counted in 0.1ml of concentrate

   C: Volume of concentrate

   V: Volume of sample in liter (it represented to total volume in one liter.

The identification of phytoplankton done by referring standard text T.V Desikachary 1959, Anand N. 1998, Prescott, G. W. 1962 Monographs, Journals and Thesis. The results were expressed as organisms per liter.

Table 4.1: Peak summer (May) and commencement of monsoon (June) 2010 Variation of Phytoplankton population count in number/litre.

Sample Station	Spot Component	May	June
BVTS1	Cyanophyceae	4020*	1890
	Chlorophyceae	3580	3500
	Bacillariophyceae	170	0
	Euglenophyceae	250	3810**
TOTAL		**8020**	**9200**
BVTS2	Cyanophyceae	80	620
	Chlorophyceae	1060	210
	Bacillariophyceae	70	2320
	Euglenophyceae	230	2000
TOTAL		**1440**	**3350**
BVTS3	Cyanophyceae	720	1350
	Chlorophyceae	790	250
	Bacillariophyceae	820	2350
	Euglenophyceae	230	80
TOTAL		**2560**	**4030**
BVTS4	Cyanophyceae	100	20
	Chlorophyceae	120	10
	Bacillariophyceae	180	630
	Euglenophyceae	0	0
TOTAL		**400**	**660**

*: *Merismopedia* sp. 3000/lit (dominant species); **: *Euglena* sp. (dominant species).

Results and Discussion

Two months such as peak summer (May) and of commencement of monsoon (June), the variation of phytoplankton species represented in Table 4.1. The total number of phytoplankton population varied from 8020 to 9200/lit at BVTS1 and 1440 to 3350/lit at BVTS2 and 2560 to 4030/lit at BVTS3 and 400 to 660/lit at BVTS4. The total phytoplankton population representing to four different groups *viz.,* Cyanophyceae, Chlorophyceae, Bacillariophyceae and Euglenophyceae. The phytoplankton diversity in lentic water bodies of Bhadravathi taluk has fluxation in number and species in the transition period of summer and onset of monsoon season.

Among the phytoplankton families in peak summer season cyanophyceae was dominant at sample station BVTS1; but in sample site BVTS2 chlorophyceae has a dominant over the cyanophyceae another two sample sites did not have specific dominance of phytoplankton members such as BVTS3 and BVTS4. In present investigation the cyanophyceae were minimum in monsoon and maximum in summer season. The following members of cyanophyceae were observed during the research period such as *Chrococcus* sp*., Oscillatoria* sp*., Nostoc* sp*., Phormidium* sp*., Anabaena* sp*., Merismopedia* sp*.*

But in onset of monsoon figures of phytoplankton population was turn over due to lower the air and water temperature and dilution of lentic water bodies by rain water. The dominance of phytoplankton population varied from station to station due their source of water from the surrounding agricultural land, domestic waste water and rain water. In sample station BVTS1 *Euglena* sp., was a dominant but chlorophyceae has a slight dominance over cyanophyceae. In sample stations BVTS2 and BVTS3 bacillariophyceae were dominance due the high dissolved oxygen, fairly good free carbon dioxide and deposition of silica due soil resin from surroundings land. In the present investigation the Bacillariophyceae varied from 170 to/lit at BVTS1 (May to June): 70 to 2320/lit at BVTS2: 820 to 2350 at BVTS3 and 180 to 630 at BVTS4 May to June).The water quality in terms levels of organic matter, dissolved oxygen, P⁰ and other physical factors play an important role in the ecological distribution of Bacillariophyceae (Sabata and Nayer, 1987).

The occurrence of chlorophyceae fairly good the total numerical density varied from sample station to stations 3580 to 3500/L at BVTS1; 1060 to 210/L at BVTS2; 790 to 250/L at BVTS3 and 120 to 10/L at BVTS4. Suresh *et al.,* 2005 reported 18 species in Chilan pond Devangere Distt., Karnataka. In the present investigation the number of chlorophyceae were observed but the dominance there were not much significance in peak summer and onset of mansoon. The following number of chlorophyceae was identified during in the investigation *viz. Pedastrum* sp.; *Scenedesmus* sp.; *Cosmarium* sp.; *Staurastrum* sp.; *Spirogyra* sp.; *Odogonium* sp.; *Costerium* sp.; *Coelcostrum* sp.; *Euastrum* sp*.* The occurrence of chlorophyceae may be due to high temperature and high nitrate and phosphate content in water bodies.

In the present investigation the seasonal numerical density of Euglenophyceae varied from 250 to 3810 at BVTS1; 230 to 200 at BVTS2; 230 to 80 at BVTS3 and none of the euglenophyceae noted in sample station BVTS4. Ammonia and dissolved organic compounds are the dominant sources of nitrogen most euglenoid algae. The occurrence of euglenophyceae least in summer season but dominant in onset of monsoon even though BVTS3 and BVTS4 did not have. The following euglenaceae were observed during the investigation *viz., Euglena polymorpha, Euglena elongata, Phacus curvicuda, Limnophylla* sp., *Phacus* sp. Pawar *et al.,* 2006 recorded 5 species of euglenophyceae at Sirur dam water. B. Munwar, 1972 reported that the high value of carbon dioxide, phosphate and low dissolved oxygen favor the grown euglenophyceae. In the cause of BVTS4 there was no occurrence of euglenophyceae may be due to high turbidity 160 NTU.

Conclusion

The present investigation of peak summer and onset monsoon have a strong influence on the distribution of phytoplankton population may be due the seasonal instability of chemical and physical parameters factors. Some of the phytoplankton community members such as *Anabaena, Chrococcus* and *Merismopedia* are as strong indicators of polluted water bodies.

References

American Public Health Association (APHA), AWWA, WEF, 1995. *Standard Methods for the Examination of Water and Wastewater,* 20th Edn. Washington, DC.

Anand, N., 1998. *Indian Freshwater Micro Algae.* Bishen Singh Mahendra Pal Singh, Dehradun.

Desikachary, T.V., 1959. *Cyanophyta.* ICAR, New Delhi.

Lackey, J.B., 1938. The manipulation and counting of river plankton and changes in some organism due to formalin preservation. *Public Health Report,* 53: 2080–2093.

Lokhanade, M.V. and Shembekar, V.S., 2009. Studies on phytoplankton diversity of Dhanegaon reservoir, Dhanegaon, Dist. Osmanabad, Maharashtra. *International Research Journal,* Vol. 2.

Murugan, C.B. and Dhande, R.R., 2000. Algal diversity with respect to pollution status of Wadali Lake, Amravathi (M.S.) India. *J.A. Biol.,* 15(1 and 2): 1–5.

Prescott, G.W., 1962. *Algae of the Western Great Lakes,* Rev. edn. Cranbrook Inst. of Sci. Bull. No. 31, Bloomfield Hills, Michigan, 977 p.

Saxena, M.M., 1990. *Environmental Analysis: Water, Soil and Air.* Agro Botanical Publishers, India.

2013, Glimpses of Animal Biodiversity   Pages **45–51**

Editor: **Dr. K. Muthuchelian,** *Vice Chancellor, Periyar University, Salem*

Published by: Daya Publishing House, NEW DELHI

Chapter 5

Psychophily and Melittophily in *Sida cordifolia* L. (Malvaceae)

A.V. Raghava Naidu, D. Jeevan Ram
and A.J. Solomon Raju

Department of Environmental Sciences,
Andhra University, Visakhapatnam – 530 003

ABSTRACT

Sida cordifolia is a seasonal shrub. Individual plants produce several solitary flowers which display their presence to flower visitors. The floral characters such as dirty white flowers, no perceptible smell, exposed nectar and spacious bottom-exposed corolla facilitate the foragers to access the nectar and pollen without any difficulty. The nectar is hexose-rich and sugar concentration ranges from 18 to 23 per cent. The flowers remain in place after the cessation of stigma receptivity; this situation may enhance attraction of flowers to the foragers. The flowers receive visits from butterflies and bees. Among butterflies, Pierids act as consistent and important pollinators while bees act as facultative pollinators. Therefore, *S. cordifolia* is both psychophilous and melittophilous. The beetle, *Mylabris phalerata* feeds on the flowers either fully or partly and such flowers do not take participation in the fruiting event. The mature fruits split apart to disperse seeds which in turn may also be driven off by high winds.

Keywords: *Sida cordifolia, Psychophily, Melittophily, Mylabris phalerata, Fruiting.*

Introduction

Sida is a genus of about 150 species of annual or perennial shrubs or herbs mostly distributed in the tropical and sub-tropical regions of the world. In India, only twelve species of this genus occur of which only seven species occur in the Eastern

Ghats forests. They include *S. cordifolia, S. cordata, S. glutinosa, S. spinosa, S. schimperiana, S. acuta* and *S. rhombifolia*. Of these, *S. acuta, S. cordata* and *S. cordifolia* are common weedy shrubs while all others are quite uncommon sub-shrubs. The present study is concerned with *S. cordifolia* only. In the study areas and in the entire length of Eastern Ghats, *S. cordifolia* grows profusely on most soil types and in nearby agricultural areas due to the use of chemical fertilizers. This chapter describes the pollination biology and seed dispersal aspects of *S. cordifolia* since there is no published information on this subject.

Materials and Methods

Sida cordifolia plants occurring at the Seshachalam Hills of southern Eastern Ghats of Andhra Pradesh were used for the study during the summer season of 2009. Twenty five tagged mature buds were followed for recording the time of anthesis and anther dehiscence; the mode of anther dehiscence was also noted by using a 10x hand lens. The details of flower morphology such as flower sex, shape, size, colour, odour, sepals, petals, stamens and ovary were described. The characteristics of pollen grains were recorded. Ten fresh flowers were used to measure the total volume of nectar/flower. The nectar sugar concentration was measured by using a Hand Sugar Refractometer (Erma, Japan) as per Dafni *et al.* (2005). Nectar analysis for sugar types was done as per the Paper Chromatography method of Harborne (1973). The stigma receptivity was tested with hydrogen peroxide from mature bud stage to flower drop as per Dafni *et al.* (2005). Regular observations were made on the insect species visiting the flowers for forage. The insects were observed for their foraging behaviour such as mode of approach, landing, probing behaviour, the type of forage they collect, contact with essential organs to result in pollination, inter-plant foraging activity in terms of cross-pollination, etc. The number of foraging visits made by each butterfly species was recorded and pooled up for calculating the percentage of visits made by each butterfly family in order to evaluate the relative importance of butterflies for pollination. Three to five specimens of each butterfly species collected at different times of the day were brought to laboratory for examining their proboscis under microscope for the presence of pollen grains in order to assess their role in pollen transfer and pollination. Fruiting ecology was also examined to the extent possible.

Results

The plant species is a weedy shrub which grows up to 1.5 meters. The stem is stout and strong. The leaves are heart-shaped, serrated and truncate. The plant thrives well during rainy season during which vegetative growth, flowering and fruiting events occur in quick succession. The flowering occurs during July and extends until late September. The flowers open during 0600-0630 h. The funnel-like corolla exposes the stigma and stamens following anthesis. The flowers are solitary, borne in axillary position, small, dirty white and bisexual. The sepals and petals are 5 each; the latter fuse with the staminal tube at the base. The staminal column is tomentose and divided into numerous stamens at the top and each bears reniform, monothecous anther. The stamens are protandrous and anther dehiscence occurs during mature bud stage by transverse slits at the top. A flower produces copious amount of pollen. The pollen grains are monads, spiny and 10.2 μm long. The ovary is 10-locular, each locule with a single ovule arranged on axile placentation; the styles are ten each terminated with a capitate stigma. The stigma attains receptivity following anthesis and continues receptivity until the evening of the second day; then onwards it shows signs of withering. The duration of receptivity was also tested by following the hydrogen peroxide test. Nectar is produced in minute amount and is collected at the base of corolla tube; it amounted to 2.7 ± 0.61 μl per flower. The nectar sugar concentration ranged from 18 per cent to 23 per cent in day 1 flowers and there is almost no trace of nectar in day 2 flowers and thereafter. The nectar contains glucose, fructose and sucrose with the first two as most dominant. The flowers remain in place for four days and fall off subsequently.

The flowers were visited during day time by butterflies for nectar collection and by other insects for nectar and pollen collection. The butterflies included 10 species representing Papilionidae, Pieridae, Nymphalidae, Lycaenidae and Hesperiidae (Figure 5.1). The Papilionidae and Hesperiidae each was represented by a single species, Nymphalidae and Lycaenidae each by 2 species and Pieridae by 4 species. The Papilionid was *Graphium agamemnon*. The Pierids were *Catopsilia pyranthe, C. pomona, Cepora nerissa* and *Eurema hecabe*. The Nymphalids were *Junonia lemonias* and *Hypolimnas missipus*. The Lycaenids were *Everus lacturnus* and *Chilades pandava*. The Hesperiid was *Borbo cinnara* (Table 5.1). The foraging activity of these butterflies except the Papilionid, gradually increased from morning to evening. This foraging trend was observed almost throughout the flowering season. *G. agamemnon* showed peak activity at 1000 h and 1500 h. The data collected on the foraging visits of butterflies of each family showed that Pierids made 40 per cent, Nymphalids 21 per cent, Lycaenids 19 per cent, Papilionids 15 per cent and Hesperiids 5 per cent of total foraging visits (Figure 5.2). The butterflies probed the flowers for nectar with great

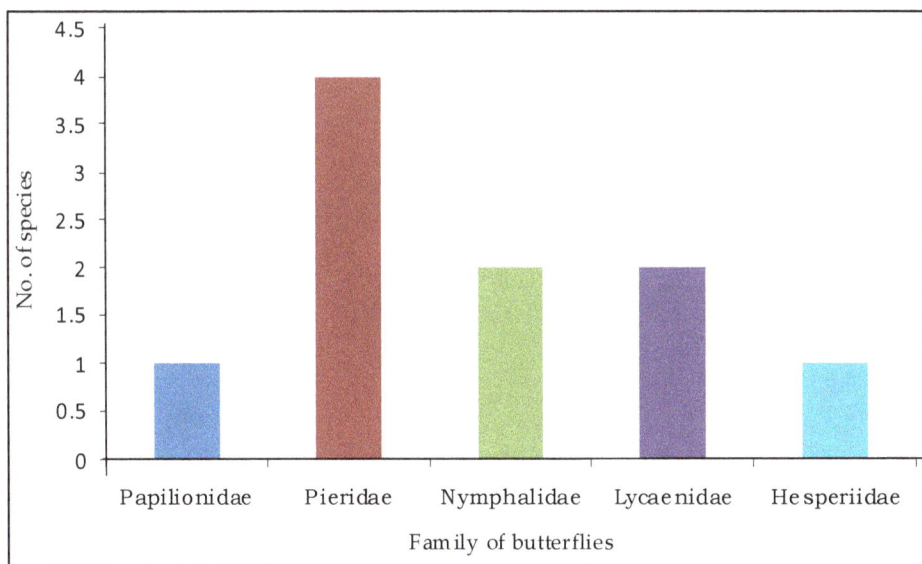

Figure 5.1: Family-wise number of butterfly species foraging for nectar on *Sida cordifolia*.

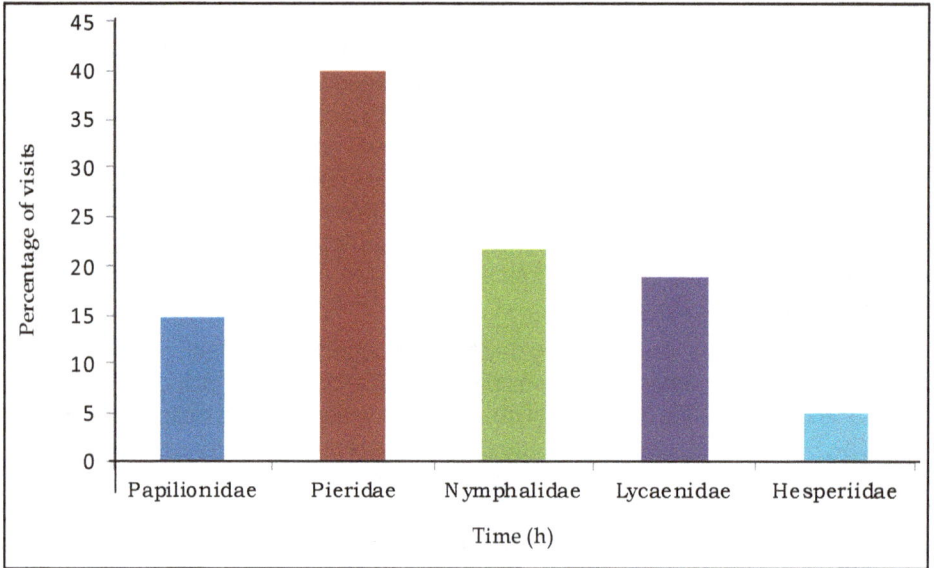

Figure 5.2: Family-wise percentage of foraging visits of butterflies on *Sida cordifolia*.

Table 5.1: List of nectar feeding insects on *Sida cordifolia*.

Family	Scientific Name	Common Name
Lepidoptera		
Papilionidae	*Graphium agamemnon*	Tailed Jay
Pieridae	*Catopsilia pomona*	Common Emigrant
	Catopsilia pyranthe	Mottled Emigrant
	Cepora nerissa	Common Gull
	Eurema hecabe	Common Grass Yellow
Nymphalidae	*Junonia lemonias*	Lemon Pansy
	Hypolimnas missipus	Danaid Egg fly
Lycaenidae	*Chilades pandava*	Plains Cupid
	Everus lacturnus	Indian Cupid
Hesperiidae	*Borbo cinnara*	Rice Swift
Hymenoptera		
Apidae	*Apis dorsata*	Rock bee
	A. cerana	Indian honey bee
	A. florea	Dwarf honey bee
	Trigona iridipennis	Stingless bee
	Ceratina viridissima	Small carpenter bee
Halictidae	*Halictus* sp.	
Eumenidae	*Eumenes conica*	

ease due to funnel-shaped corolla with exposed flower base. They never attempted to manipulate the mature buds for nectar. The flowers provide comfortable landing place for butterflies to probe the flowers for nectar. The solitary state of flowers at plant level was considered to be uneconomical for butterflies to collect nectar in relation to other nectar sources reported in this section. A sample of 3-5 specimens of all butterfly species was used to examine the pollen carrying capacity of their proboscides. The results indicated that the proboscides contained pollen grains ranging from 34-142 in Papilionids, 78-219 in Pierids, 43-178 in Nymphalids, 34-218 in Lycaenids and 34-79 in Hesperiids. All these butterflies while collecting nectar contacted the stamens and stigmas during which pollination occurred. The butterflies made visits to collect nectar from different plants which show patchy distribution; such a foraging behaviour could promote cross-pollination.

Other insects included bees, *A. dorsata, A. cerana, A. florea, Trigona iridipennis, Ceratina viridissima* (Apidae), *Halictus* sp. (Halictidae), the wasp, *Eumenes conica* (Eumenidae) (Table 5.1) and *Mylabris phalerata* (Meloidae). All these except this last species approached the flowers in upright position, landed on the corolla and probed the flowers for the forage. All bee species collected both pollen and nectar in the same foraging visit while the wasp collected only nectar. While collecting the forage, both bees and the wasp contacted the stamens and stigmas due to which pollination occurs. The bees were consistent in their foraging activity throughout the flowering season while the wasp was an occasional nectar feeder. At the study site, *T. iridipennis* exhibited fidelity to *S. cordifolia* and collected its pollen voraciously until the floral source was exhausted. A nest of this bee species nearby was examined for pollen source in pollen pots. It showed that all pollen pots were filled with this pollen only. The beetle, *M. phalerata* fed on the flowers, especially the corolla, stamens and stigma. A sample of 125 flowers showed that 37.5 per cent of them were partly or fully fed by this beetle.

The fruit growth and development begins immediately after pollination and fertilization. A sample of 175 flowers showed 31 per cent fruit set in open-pollinations. The fruits are small, 6-8 mm diameter. Each fruit produces 3-6 seeds; the fruits dehisce when mature; maturation of fruits occurs within a month. The mature fruits split apart to disperse seeds. This occurs during September-December. The seeds brown to black, smooth, obovoid with three angles, 2mm long, minutely pubescent around the raphe; they germinate to produce new plants in the next year following monsoonal rains.

Discussion

Croat (1978) reported that most species of Malvaceae are pollinated by pollen feeding insects. He also mentioned that *Hibiscus rosa-sinensis* is pollinated by Papilionid butterflies in the entire Asia. Solomon Raju (2007) reported on the pollinators of a few members of Malvaceae. Butterflies and pollen feeding bees are pollinators of *Abutilon indicum* while bees are the exclusive pollinators of the endemic *A. ranadei, Hibiscus vitifolius* and *Pavonia zeylanica*. Croat (1978) stated that *Sida rhombifolia* is self-pollinated but are also pollinated by the beetle *Dysdercus cingularis* and by numerous bees. Taylor (2002) stated *S. fallax* is probably anemophilous. Solomon

Raju (2007) suggested that butterflies and small pollen feeding bees effect pollination in *S. acuta*.

In *S. cordifolia*, the vegetative and reproductive phases occur in quick succession during rainy season. Each individual plant with several solitary conspicuous flowers borne in axillary position stands out visually and attracts butterflies and bees. The floral characters such as dirty white flowers, no perceptible smell, exposed nectar and spacious corolla facilitate the foragers to access the nectar and pollen without any difficulty. Since the anthers dehisce during mature bud stage, the newly open flowers offer both nectar and pollen readily to the foragers. The morning anthesis is an important trait for the plant to present floral rewards right from morning to the prospective pollinators. The nectar is hexose-rich and sugar concentration ranges from 18 to 23 per cent. This is in agreement with the generalization that Verbenaceae nectars are characteristically hexose-rich (Galetto and Bernardello 2003) and that the sugar concentration range recorded optimizes the net energy gain by the visiting butterflies (Kingsolver and Daniel 1979). Butterflies and bees begin their foraging activity as soon as the flowers are open. They collect the nectar with great ease; the bees also collect pollen in the same foraging visit. At the same time, both of them contact the anthers and stigma and pollinate the flowers. The intense foraging activity of both butterflies and bees throughout the day surely effects pollination in day 1 flowers. The day 2 flowers are mostly devoid of nectar and pollen and hence are not profitable for both butterflies and bees. However, the visits of pollen carrying butterflies and bees to these flowers effect pollination until evening of the 2nd day since the stigma is receptive until that time. The presence of pollen grains on the proboscis of butterflies indicates that they are pollen carriers and effect pollination while collecting nectar. Since the flowers produce copious pollen, the pollen collection by bees may not drastically reduce the pollen availability for pollination. However, in case of the stingless bee, *Trigona iridipennis*, it is a voracious pollen feeder and consistently collects pollen until the resource is exhausted. The pollen pots of this bee in the nest also attest this observation. In effect, its intense pollen collection activity may affect the pollination rate in *S. cordifolia*. The retention of flowers after the cessation of stigma receptivity may enhance attraction of the floral source to the visiting foragers. Since each plant offers a limited number of flowers each day, the butterflies and bees move between plants in search of the forage and in the process effect cross-pollination also. The pollen grains that reach the capitate stigma germinate quickly and produce pollen tubes in a polysiphonous manner. They penetrate through the transmitting tissue of the style and enter ovule in a porogamous manner; subsequently fertilization occurs (Venkata Rao 1955). *S. cordifolia* flowers attract the butterflies of all Lepidopteran families but Pierids seem to be relatively more attracted than other butterflies. The butterflies collect nectar throughout the day but their foraging activity pattern is directly related to the standing nectar crop at different times of the day. A wasp, *Eumenes conica* also utilizes *S. cordifolia* as a nectar source; it also effects pollination but it is not a consistent and frequent nectar forager, and hence is an opportunist rather than a pollinator. Therefore, *S. cordifolia* is both psychophilous and melittophilous.

The literature does not give any clue whether the flowers of *Sida* species serve as a source of food for beetles. In *S. cordifolia*, the beetle, *Mylabris phalerata* feeds on the flowers either fully or partly and such flowers do not take participation in the fruiting event. The flower predation rate is quite visible and hence reduces the success rate of sexual reproduction to a considerable extent. But, the flower predation by *M. phalerata* is probably a natural control to limit the spread of *S. cordifolia* since it grows as a weed and menace for the survival of many other low ground herbs.

In *S. cordifolia*, the natural fruit set rate is an indication that it is both self–and cross-pollinating; it may increase further if the flowers attract more species of butterflies and discourage pollen collecting bees. The mature fruits with up to six seeds split apart and disperse seeds which in turn may also be carried away by high wind. The seeds thus dispersed remain in the soil until the next rainy season to produce new plants. Therefore, *S. cordifolia* disperses seeds itself and the wind may also effect seed dispersal.

References

Croat, T.B., 1978. *Flora of Barro Colorado Island.* Stanford Univ. Press, California.

Dafni, A., Kevan, P.G. and Husband, B.C., 2005. *Practical Pollination Biology.* Enviroquest Ltd., Cambridge.

Galetto, L. and Bernardello, G., 2003. Nectar sugar composition in angiosperms from Chaco and Patagonia (Argentina): An animal visitor's matter? *Plant Syst. Evol.,* 238: 69–86.

Harborne, J.B., 1973. *Phytochemical Methods.* Chapman and Hall, London.

Kingsolver, J.G. and Daniel, T.L., 1979. On the mechanics and energetics of nectar feeding in butterflies. *J. Theor. Biol.,* 76: 167–179.

Rao, Venkata, 1955. Embryological studies of Malvaceae II. *Proc. Nat. Inst. Sci. Ind.,* 21B: 53–67.

Solomon Raju, A.J., 2007. *Indian Plant Reproductive Ecology and Biodiversity.* Today and Tomorrow's Printers and Publishers, New Delhi.

Taylor, S.A., 2002. *Sida fallax* Walp. Ilima, pp.1–2.

2013, Glimpses of Animal Biodiversity   *Pages* **52–59**
Editor: **Dr. K. Muthuchelian,** *Vice Chancellor, Periyar University, Salem*
Published by: **Daya Publishing House, NEW DELHI**

Chapter 6

Psychophily and Melittophily in *Vitex altissima* L.F. (Verbenaceae)

*K. Venkata Ramana and A.J. Solomon Raju**

*Department of Environmental Sciences,
Andhra University, Visakhapatnam – 530 003*

ABSTRACT

Vitex altissima is a deciduous tree species. In *V. altissima*, the floral characteristics such as purple flowers, no perceptible smell, tubular corolla, and non-exposed nectar deposited at the bottom of the corolla tube strongly select the visitors and pollinators, demanding a more elaborated intrafloral behaviour. The lower elaborate lower lip of the corolla provides comfortable landing place for the foragers. Since the anthers dehisce during mature bud stage, the newly open flowers offer both nectar and pollen readily to the foragers. The nectar is hexose-rich and sugar concentration ranges from 24-28 per cent. Nymphalid and Lycaenid butterflies and bees pollinate the flowers. Although bees effect pollination, they reduce the availability of pollen for pollination. Therefore, *V. altissima* is primarily psychophilous while melittophily is an additional pollination mode. The Scarabaeid, *Popillia impressipyga* acts as bud and flower predator; its feeding affects the reproductive success of the plant. The calyx in fertilized flowers gradually enlarges and acts as fruiting calyx to harbour the growing seeds. The seeds mature within six weeks and fall off to the ground due to gradual wrinkling of the drying calyx from the base to the top. Since seeds are light in weight, high winds disperse them to other places when dry conditions exist due to dry spell that usually occurs during rainy season.

Keywords: *Vitex altissima, Psychophily, Melittophily.*

___

* Corresponding Author: E-mail: ajsraju@yahoo.com

Introduction

The genus *Vitex* is comprised of about 300 species almost all occurring in the tropics. Most of the species of this genus are common in swamps and primary rainforests, and are often prominent in secondary, coastal or fire-influenced vegetation or the drier evergreen forests (Whitten *et al.,* 1996) and the monsoon forest in the Philippines (Oldfield *et al.,* 1998). In the present study, the deciduous monsoon forest in the Eastern Ghats harbours four species, *V. negundo, V. peduncularis, V. leucoxylon* and *V. altissima;* the first one is a deciduous shrub while all other species are deciduous trees. *V. negundo* and *V. altissima* are common while the other two species are rare (Chetty *et al.,* 2008). *V. altissima* which shows small population at the present study site was investigated for its pollination biology and seed dispersal since there is no information on its reproductive biology despite its importance in the places of its occurrence in the Eastern Ghats of Andhra Pradesh.

Materials and Methods

Vitex altissima trees occurring at the Seshachalam Hills of southern Eastern Ghats of Andhra Pradesh were used for the study during the summer season of 2009. Twenty five tagged mature buds were followed for recording the time of anthesis and anther dehiscence; the mode of anther dehiscence was also noted by using a 10x hand lens. The details of flower morphology such as flower sex, shape, size, colour, odour, sepals, petals, stamens and ovary were described. Ten mature but undehisced anthers of staminate flowers were collected from five different plants and placed in a Petri dish. Later, each time a single anther was taken out and placed on a clean microscope slide (75 x 25 mm) and dabbed with a needle in a drop of lactophenol-aniline-blue. The anther tissue was then observed under the microscope for pollen, if any, and if pollen grains were not there, the tissue was removed from the slide. The pollen mass was drawn into a band, and the total number of pollen grains was counted under a compound microscope (40x objective, 10x eye piece). This procedure was followed for counting the number of pollen grains in each anther collected. Based on these counts, the mean number of pollen produced per anther was determined. The mean pollen output per anther was multiplied by the number of anthers of a flower for obtaining the mean number of pollen grains per flower. The characteristics of pollen grains were also recorded. Ten fresh flowers were used to measure the total volume of nectar/flower. The nectar sugar concentration was measured by using a Hand Sugar Refractometer (Erma, Japan) as per Dafni *et al.* (2005). Nectar analysis for sugar types was done as per the Paper Chromatography method of Harborne (1973). The stigma receptivity was tested with hydrogen peroxide from mature bud stage to flower drop as per Dafni *et al.* (2005). Regular observations were made on the insect species visiting the flowers for forage. The insects were observed for their foraging behaviour such as mode of approach, landing, probing behaviour, the type of forage they collect, contact with essential organs to result in pollination, inter-plant foraging activity in terms of cross-pollination, etc. Three to five specimens of each butterfly species collected at different times of the day were brought to laboratory for examining their proboscis under microscope for the presence of pollen grains in order to assess their role in pollen transfer and pollination. Fruiting ecology was also examined to the extent possible.

Results

Plant Phenology and Flower Morphology

It is a medium-sized deciduous tree which grows up to 25-33 m. The leaves are tri-foliate with slight pubescence. It is leafless during summer season from March to May. Leaf flushing occurs during late May and June. The flowering occurs during late June to late July; it is almost synchronous at population level. The inflorescences are paniculate cymes and borne in leaf axils and also terminally, and are minutely pubescent. The flowers are small, odourless, bisexual and zygomorphic. The calyx is cupular 3 mm long, fulvous-villous with five 1 mm long oblong lobes. The corolla is tubular with villous throat, whitish purple to violet, bi-lipped; the upper lip is 2.5 mm long with two triangular sub-acute lobes while the lower lip is 4 mm long, 3-lobed with the middle lobe more than twice as large as lateral ones. The stamens are four, didynamous, epipetalous and the anther locules attached only at the tip becoming divaricate; the filaments are hairy at the base. The ovary is bicarpellary, tetralocular with a total of four ovules arranged on axile placentation. The style filiform, curved and terminates with bifid stigma.

Floral Biology

The flowers are open during 0600-0800 h and anthers dehisce by longitudinally during mature bud stage. The pollen output per flower is 2,567 ± 32.8 pollen grains. The pollen grains are tricolpate, 30 ± 1.90 μm long on polar axis and 22 ± 1.80 μm long on equatorial axis with tapering ends, membrane granular, exine tectate, smooth to granulate and perforate. The stigmatic lobes unfold after anthesis indicating the commencement of the receptivity and continue receptivity until the noon of the next day after which the stigma shows the signs of withering. Hydrogen peroxide test was also used to confirm the stigma receptivity and it showed almost the same time schedule of stigma receptivity. The nectar is produced in mature bud stage and its production ceases after anthesis. A flower produces an average of 4.1 ± 0.23 μl of hexose-rich nectar with a sugar concentration of 24-28 per cent; glucose and fructose sugars were equally dominant while sucrose was present in traces. The flowers fall off in the evening of the next day.

Flower Visitors and Pollination

The flowers were visited during day time by butterflies for nectar collection and by bees for nectar and pollen collection. The butterflies included 8 species representing Nymphalidae and Lycaenidae. The Nymphalidae was represented by 6 species while Lycanidae by 2 species (Figure 6.3). The Nymphalids were *Junonia lemonias, Euploea core, Phalanta phalantha, Tirumala limniace, Parantica aglea* and *Neptis hylas*. The Lycaenids were *Jamides celeno* and *Chilades pandava*. Of these, the individuals of Nymphalid butterflies were more than the Lycaenids at the flowers throughout the flowering season. The foraging activity of these butterflies showed an increase in activity is followed by a decrease in activity both in the forenoon and afternoon period (Figures 6.1 and 6.2). This foraging trend was observed almost throughout the flowering season. The data collected on the foraging visits of butterflies showed that Nymphalids made 77.7 per cent, while Lycaenids 22.3 per cent of total visits (Figure 6.4). The

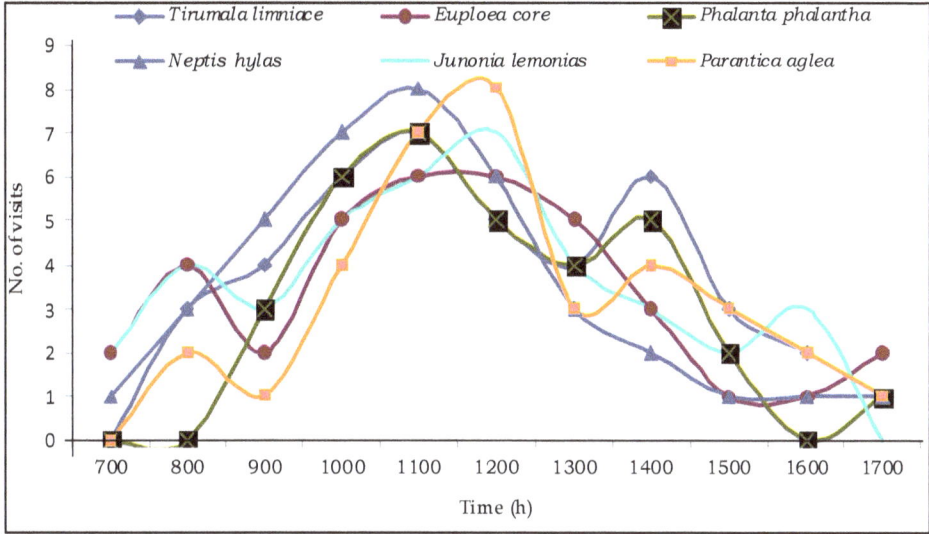

Figure 6.1: Hourly nectar foraging activity of Nymphalid butterflies on *Vitex altissima*.

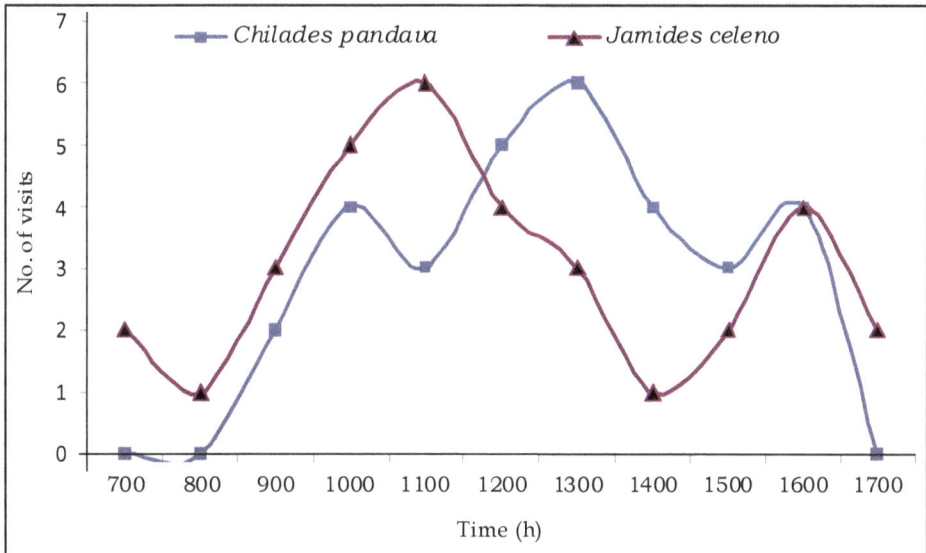

Figure 6.2: Hourly nectar foraging activity of Lycaenid butterflies on *Vitex altissima*.

butterflies with varying proboscis lengths were found to access the nectar in the tubular flowers. They never attempted to manipulate the mature buds for nectar. The small clustered flowers borne in paniculate cymes were found to be energetically economical to probe several flowers in each visit in succession for nectar by butterflies before their departure. The butterflies held the inflorescence axis/adjacent flowers with their fore and hind legs to probe the flowers for nectar. A sample of 3-5 specimens of all butterfly species was used to examine the pollen carrying capacity of their

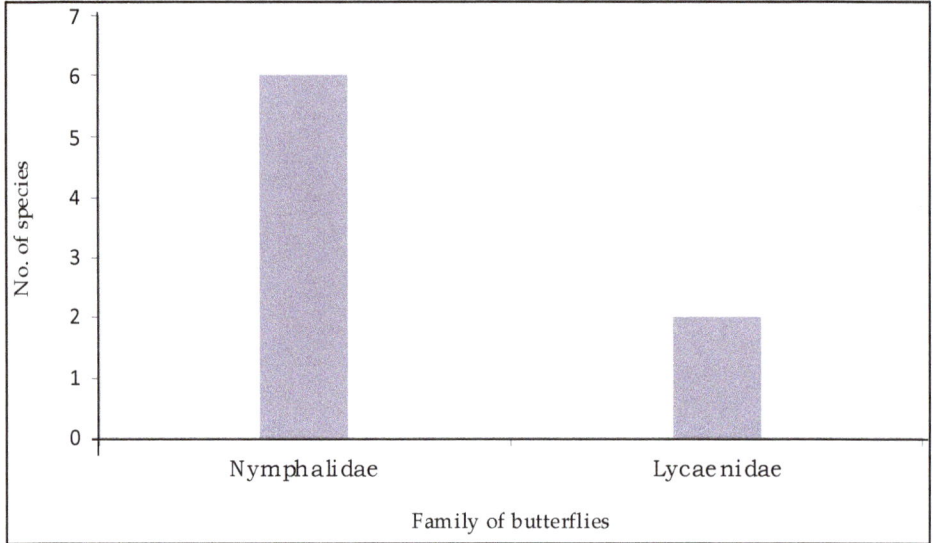

Figure 6.3: Family-wise number of butterfly species foraging for nector on *Vitex altissima*.

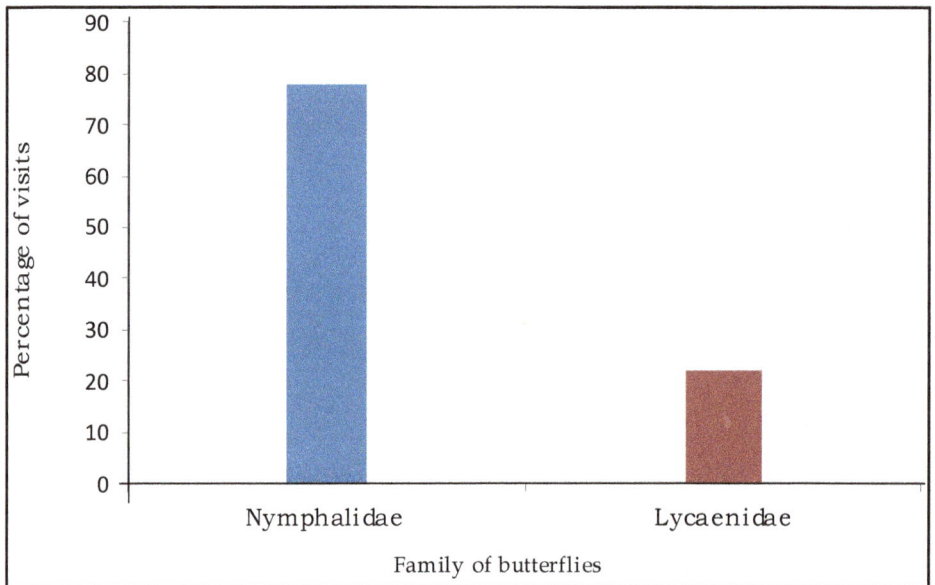

Figure 6.4: Family-wise percentage of foraging visits of butterflies on *Vitex altissima*.

proboscides. The results indicated that the proboscides invariably contained pollen grains ranging from 47-189 in Nymphalids and 27-49 in Lycaenids. The butterflies stretched out their proboscis to reach the floral base to access nectar; while doing so the proboscis invariably contacts the stigmatic lobes which extend beyond the length of stamens and hence effects pollination. The butterflies frequently moved between individual plants of *V. altissima* which occur scattered in the forest; this inter-tree foraging activity was considered to be important in promoting cross-pollination. The study showed that Nymphalids play an important role in the pollination of *V. altissima.*

Bee foragers included *A. dorsata, A. cerana, A. florea* (Apidae) and *Halictus* sp. (Halictidae). They approached the flowers always in upright position, landed on the spacious lower lip prior to probing. As soon as they landed, they inserted their tongue into the flower base to collect nectar during which they contacted the bifid stigma and dehisced anthers forcibly; this intra-floral behaviour surely results in pollination. They also collected pollen prior to their departure from the flower. Further, they collected both nectar and pollen from several flowers before they departed from the flowering inflorescence. The bees were consistent in their foraging activity until noon throughout the flowering season of *V. altissima.* The pollen collection activity of bees was considered to be reducing the availability of pollen for pollination to some extent.

A Scarabaeid species, *Popillia impressipyga* was also found to be feeding on the buds and flowers voraciously. A sample of 500 buds and 750 flowers was observed to record the percentage of predation by this species. The study showed that 37.8 per cent of the buds and 63 per cent of flowers were damaged. This predator was found to be impacting the success rate of sexual reproduction.

Fruiting Ecology

The fruit growth and development begins immediately after pollination and fertilization. A sample of ten tagged terminal inflorescences consisting of 467 flowers showed 76 per cent fruit set in open-pollinations. The fruits mature within one and half a month; they are small and light purple initially and then turn dark purplish brown. Each fruit produces 2-4 seeds and are housed in fruiting calyx. Fruit is a drupe subtended by enlarged calyx, ovoid, smooth, 5-8 mm diameter with four purplish black hard seeds. The mature seeds fall off from the fruiting calyx due to its gradual withering and shrinking from the base to tip. Seed dispersal occurs in September-October. The fallen seeds germinate during the next rainy season to produce new plants.

Discussion

Reddy *et al.* (1992) reported that *V. negundo* flowers attract different diurnal insects but they have not mentioned which species visit them. However, these authors stated that two wasp species, *Rhynchium metallicum* and *Ropalidia* sp. puncture the corolla tube in mature bud stage to collect nectar and almost all other insect species use the punctured part to collect nectar by bypassing the pollination apparatus. The perforated flowers set the lowest fruit set. Casual observations on *V. negundo* at the study sites showed that the flowers are primarily pollinated by butterflies which always probe the flowers from the front and perforation by these wasps has been found to be absent although they occur in the habitat during the flowering phase. In *V. altissima,* the floral characteristics are almost similar to *V. negundo*–flower size, colour, corolla shape, symmetry and sex. A suite of floral characteristics such as purple flowers, no perceptible smell, tubular corolla, and non-exposed nectar deposited at the bottom of the corolla tube strongly select the visitors and pollinators, demanding a more elaborated intrafloral behaviour. The lower elaborate lower lip of the corolla provides comfortable landing place for the foragers. Since the anthers dehisce during mature bud stage, the newly open flowers offer both nectar and pollen readily to the foragers. The nectar is hexose-rich and sugar concentration ranges from 24-28 per cent. This is in agreement with the generalization that Verbenaceae nectars are characteristically hexose-rich (Galetto and Bernardello 2003) and that the sugar concentration range recorded optimizes the net energy gain by the visiting butterflies (Kingsolver and Daniel 1979). Butterflies and bees begin their foraging activity as soon as the flowers are open. They reach the bottom of the corolla and take the nectar; the bees also collect pollen in the same foraging visit. At the same time, both of them contact the anthers and stigma and pollinate the flowers. The intense foraging activity of both butterflies and bees throughout the day surely effects pollination in day 1 flowers. The day 2 flowers are mostly devoid of nectar and pollen and hence are not profitable for both bees and butterflies. However, the visits of pollen carrying butterflies and bees to these flowers effect pollination until noon since the stigma is receptive until that time. The presence of pollen grains on the proboscis of butterflies indicates that they are pollen carriers and effect pollination while collecting nectar. *V. altissima* flowers attract only Nymphalid and Lycaenid butterflies; the former category is very important for pollination. Among Nymphalids, *Tirumala limniace* is the most prominent pollinator. Since bees collect pollen voraciously, they reduce the availability of pollen for pollination although they themselves bring about pollination during forage collection. Therefore, *V. altissima* is primarily psychophilous while melittophily is an additional pollination mode. Barbola *et al.* (2006) mentioned that psychophily and melittophily appear to be usual in Verbenaceae to which *V. altissima* belongs. Psychophily and melittophily are reported in *Stachytarpheta mutabilis* (Stone *et al.*, 1988), *S. glabra* (Antonini *et al.*, 2005) and *S. maximiliani* (Barbola *et al.*, 2006). *Lantana camara* and *L. trifolia* is exclusively psychophilous (Schemske 1976). Later, Bhattacharya and Mandal (1998) reported that *L. camara* is also entomophilous. Further, they also reported that *Tectona grandis* and *Vitex negundo* are both psychophilous and entomophilous. Hopper (1980) mentioned that *Syzygium tierneyanum* is psychophilous. These various reports suggest that both psychophily and entomophily are prevalent in the members of Verbenaceae.

In *V. altissima*, the buds and flowers being very delicate and transparent attract the Scarabaeid species, *Popillia impressipyga*. Numerous individuals of this species feed simultaneously on each individual plant. This feeding activity is severely impacting the success of sexual reproduction which is the only mode for its natural propagation.

In *V. altissima*, the natural fruit set evidenced in tagged flowers indicates that it is both self- and cross-pollinating. The calyx gradually enlarges and acts as fruiting calyx to harbour the growing seeds. The seeds mature within six weeks and fall off to the ground due to gradual wrinkling of the drying calyx from the base to the top. Since seeds are light in weight, high winds disperse them to other places when dry conditions exist due to dry spell that usually occurs during rainy season. This led Sundarapandian *et al.* (2005) to state that *V. altissima* is anemochorous. The present study suggests that rainy season is not favourable for seed dispersal by wind and hence *V. altissima* is not adapted for anemochory. Therefore, *V. altissima* is specialized in that it disperses its seeds through the fruiting calyx.

References

Antonini, Y., Souza, H.G., Jacobi, C.M. and Mury, F.B., 2005. Diversidade e comportamento dos insetos visitants florais de *Stachytarpheta glabra* Cham. (Verbenaceae), em uma area de campo ferruginoso, Ouro Preto, MG. *Neotrop. Entomol.*, 34: 555–564.

Barbola, F.I., Laroca, S., Almeida, M.C. and Nascimento, E.A., 2006. Floral biology of *Stachytarpheta maximiliani* Scham. (Verbeneaceae) and its floral visitors. *Rev. Bras. Entomol.*, 50: 1–14.

Bhattacharya, A. and Mandal, S., 1998. A contribution to the diversity of insects with reference to pollination mechanism in some angiosperms. In: *Proc. National Seminar on Environmental Biology*, Visva-Bharati University, Santiniketan.

Chetty, K.M., Sivaji, K. and Tulasi Rao, K., 2008. *Flowering Plants of Chittoor District, Andhra Pradesh, India*. Students Offset, Tirupati.

Dafni, A., Kevan, P.G. and Husband, B.C., 2005. *Practical Pollination Biology*. Enviroquest Ltd., Cambridge.

Galetto, L. and Bernardello, G., 2003. Nectar sugar composition in angiosperms from Chaco and Patagonia (Argentina): An animal visitor's matter? *Plant Syst. Evol.*, 238: 69–86.

Harborne, J.B., 1973. *Phytochemical Methods*. Chapman and Hall, London.

Hopper, S.D., 1980. Pollination of the rain-forest tree *Syzygium tierneyanum* (Myrtaceae) at Kuranda, Northern Queensland. *Aust. J. Bot.*, 28: 223.

Kingsolver, J.G. and Daniel, T.L., 1979. On the mechanics and energetics of nectar feeding in butterflies. *J. Theor. Biol.*, 76: 167–179.

Oldfield, S., Lusty, C. and MacKinven, A., 1998. *The World List of Threatened Trees*. World Conservation Press, Cambridge, UK.

Reddy, T.B., Rangaiah, K., Reddi, E.U.B. and Subba Reddi, C., 1992. Consequences of nectar robbing in the pollination ecology of *Vitex negundo* (Verbenaceae). *Curr. Sci.*, 62: 690–691.

Schemske, D.W., 1976. Pollinator specificity in *Lantana camara* and *L. trifolia* (Verbenaceae). *Biotropica* 8: 260–264.

Stone, G.N., Amos, J.N., Stone, T.F., Knight, R.L., Gay, H. and Parrott, F., 1988. Thermal effects on activity patterns and behavioural switching in a concourse of foragers on *Stachytarpheta mutabilis* (Verbenaceae) in Papua New Guinea. *Oecologia* (Berlin), 77: 56–73.

Sundarapandian, S.M., Chandrasekaran, S. and Swamy, P.S., 2005. Phenological behaviour of selected tree species in tropical forests at Kodayar in the Western Ghats, Tamil Nadu, India. *Curr. Sci.*, 88: 805–810.

Whitten, T., Soeriaatmadja, R.E. and Afiff, S.A., 1996. The ecology of Java and Bali. In: *The Ecology of Indonesia Series*, 2 Periplus, Singapore.

2013, Glimpses of Animal Biodiversity   *Pages* **60–80**

Editor: **Dr. K. Muthuchelian,** *Vice Chancellor, Periyar University, Salem*

Published by: **Daya Publishing House, NEW DELHI**

Chapter 7

Identification and Taxonomical Characterization of Potential Polyhydroxyalkanoate Accumulating Strains Isolated from Madurai District, Tamil Nadu, India

K. Sujatha[1], V. Gangadevi[1], A. Mahalakshmi[1], P. Gunasekaran[1]* and R. Shenbagarathai[2]

[1]*Networking Centre in Biological Sciences, School of Biological Sciences, Madurai Kamaraj University, Madurai*
[2]*PG Department of Zoology and Biotechnology, Lady Doak College, Madurai*

ABSTRACT

Polyhydroxyalkanoates (PHAs) represent a large class of microbial polyesters, which are widely distributed in prokaryotes and these have gained a considerable interest due to their potential application as biodegradable, renewable and environmentally friendly properties. Acceptance for large scale production of bacterial polyesters will therefore be determined by the following operational factors: (i) fulfilling an urgent market need (ii) installation of efficient composting systems and (iii) competing with a leading advantage over synthetic

* Corresponding Author: E-mail: gunagenomics@gmail.com

plastics in terms of quality, processing performance and meeting the requirements for registration as food packages. Hence, in this work much effort has been devoted to reduce the cost of PHA by isolating better indigenous bacterial strains. Among the 600 bacterial strains examined, only 2 isolates (LDC-5 and LDC-25) showed better accumulation of PHA. Biochemical and 16S rRNA sequence analysis demonstrated that strain LDC-5 is positioned with the type strain of *Pseudomonas putida* (L28676-97 per cent), whereas LDC-25 phylogenetically is closer to *Pseudomonas fulva* (DQ141541-97 per cent) strain. 16S rRNA secondary structure prediction revealed the presence of potential helical structure in LDC-5 and LDC-25. The spectral analysis demonstrated the presence of PHA specific ester carbonyl band at 1735 cm^{-1}. Therefore it is hoped that the results described in this work will be useful for the enhanced production of technically interesting biodegradable polymer in near future.

Introduction

The past hundred and fifty years of this millennium has witnessed an unbelievable pace in scientific and technological development making this century appear like a sci-fi novel. Plastic is considered to be an invaluable gift of modern science and technology to mankind. This unique wonder material has some mutually exclusive qualities of being very light, yet strong and economical. The mammoth scale of use of plastics and their improper disposal has threatened the natural environment the world over. Thus the need of the hour is to find alternative materials, having the physical and industrial qualities of petrochemically derived plastics, yet being biodegradable (Atlas, 1993). Biodegradable polymers have evolved significantly over the past fifteen years. Polyhydroxyalkanoates (PHAs) are examples of such biopolymers being synthesized as carbon and energy storage products by a broad range of bacteria. PHAs are of technological and commercial interest, because the extracted materials are thermoplastics, which can be processed into a variety of consumer goods and medical devices. In contrast to petroleum-based plastics, these biologically produced polymers are synthesized from renewable resources and are completely biodegradable (Doi, 1990; Anderson and Dawes, 1990).

PHAs are classified into three types: Short Chain Length (SCL) PHA, Medium-Chain Length (MCL) PHA and Short Chain Length-Medium Chain Length PHA (SCL-MCL) based on its monomer composition. Bacterium *Cupriavidus necator* (formerly known as *Wautersia eutropha* and earlier as *Ralstonia eutropha*), *Comamonas acidovorans* ATCC 15668, *Pseudomonas glathei* ATCC 29195 and *Paracoccus versutus* (environmental isolated strain) accumulates SCL-PHA group and the monomers produced by this group range from 3 to 5 carbon atoms (Wu *et al.*, 2003); whereas MCL-PHA group is produced by Pseudomonads (*e.g. Pseudomonas putida* ATCC 12633, *Pseudomonas putida* GPo1 (formerly known as *Pseudomonas oleovorans* GPo1 ATCC 29347 (Van Beilen *et al.*, 2001) and *Pseudomonas citronellolis* ATCC 13674) contains monomers ranging from 6 to 16 carbons in length (Steinbuchel, 1991). *Pseudomonas* sp.61-3, *Pseudomonas oleovorans* strain B-778 and *Pseudomonas stutzeri* are found to produce a mixture of PHB and MCL-PHA (Matsusaki *et al.*, 1998; Ashby *et al.*, 2002; Chen *et al.*, 2006). It is the diverse hydroxyalkanoate monomers produced by various microorganisms that attract significant interest of microbiologists as well as polymer scientists (Kung *et al.*, 2007). PHAs producers, though detected in many natural environments–estuarine and intertidal sediments, ground water aquifers, sewage sludge, gypsum rich sands, rivers, root nodules have lent themselves to PHA accumulating indigenous bacterial populations of natural ecosystems (Rothermich *et al.*, 2000). The major barrier to commercialization of PHA has been, and is, its price as compared to conventional petrochemical based plastic materials. Hence, alternative strategies to boost PHA production are imperative. The most viable strategy is: identification of potent indigenous PHA producers with high PHA accumulating capacity. There are many traditional first-line screening methods for PHA accumulating bacteria. These include both phenotypic and genotypic detection methods (Spiekermann *et al.*, 1999; Sheu *et al.*, 2000; Solaiman *et al.*, 2000; Shamala *et al.*, 2003; Solaiman and Ashby, 2005).

Two potential strains (LDC-5 and LDC-25) were selected and characterized at their molecular level using PCR analysis. Preliminary PCR analysis positioned the two strains within the genus *Pseudomonas.* These strains were further characterized for their species level identification, but precise identification of these bacterial strains up to their species level is not an easy task using conventional biochemical techniques. In microbiology, conventional identification techniques are based on culturing from soil/water samples and subsequently subjecting grown cultures to Gram staining and plating on selective culture plates to observe colony morphology, after which conventional biochemical identification tests are performed. A major drawback in this approach is long delay, as growth detection by culture systems takes approximately two days and subsequent identification requires one or two additional overnight incubations. To get around the pitfalls of these conventional methods, a new standard for identifying bacteria based on 16S rDNA amplification has begun to evolve as a good alternative. Candidates included in this taxonomic analysis are bacterial genes that code for the 5S, 16S, 23S rRNA and the spaces between the genes. The part of DNA most commonly used for taxonomic purposes for bacteria is the 16S rRNA gene (Bottger, 1989; Garrity and Holt, 2001; Harmsen and Karch, 2004). The 16S rRNA gene is also designated 16S rDNA (the term is used interchangeably) and according to the current ASM (American Society for Microbiology) policy usage the name is "16S rRNA gene" (Clarridge, 2004). In general, the comparison of the 16S rRNA gene sequences allows differentiation between organisms at the genus level across all major phyla of bacteria, in addition to classifying strains at multiple levels, including the species and subspecies level. The 16S rRNA gene comparisons showed that the phylogeny of microbes corresponded well with their taxonomy on the basis of physiological and bio-chemical characteristics (Devereux *et al.*, 1984; Woese, 1987; Hugenholtz *et al.*, 1998; Baker *et al.*, 1999). Hence, in this study, 16S rRNA sequence analysis has been used as a powerful tool to clarify the taxonomic affinities of a wide range of taxa. Also, spectral studies supported that the polymer obtained from the selected strain is an environmentally important one that meet commercial interest.

Materials and Methods

Micro-organism and Culture Conditions

The PHA accumulating strains LDC-5 and LDC-25 were grown aerobically at 37°C in 250 ml Erlenmeyer flasks containing 50 ml of sterilized Nutrient Medium (NB), pH 7.0 ± 0.2 at 150 rpm. *Pseudomonas oleovorans* 14682 and *Pseudomonas corrugata* 388 were used as positive controls, *E. coli* JM109 and Qiagen EZ competent cells as negative control. LB broth (Hi media) was used for general cultivation of bacteria and 1.5 per cent Bacto agar was added whenever solid media was required. The cultures were incubated at 37°C for 48 hrs.

Method of Mega Plasmid DNA Isolation

Mega plasmid DNA was isolated from bacterial strains following the methods of Jensen *et al.*, 1995. The bacterial colonies were selected and each was inoculated into 2 ml LB containing the appropriate antibiotic. The cells were spun down by a quick spin and the cell pellet was resuspended in 100 μl E buffer [(g/100ml: Sucrose: 15g; EDTA (0.5M pH.8.0): 40ml; Tris (1M pH 8.0): 2ml; double distilled water to the volume]. 200 μl of lysis buffer [g/100ml: SDS: 3.0g; Tris (1M pH 12.5): 0.5ml; double distilled water to the volume] and the tube was shaken well and kept at 60°C for 30 min. 5U of proteinase K (50μg/ml) was added and they were mixed well by hand shaking 20 times and kept at 37°C for 90 min. Finally 1ml of Phenol: Chloroform: Isoamyl alcohol (500:480:20) was added, the contents were mixed 40 times and spun at 10,000 g for 3 min. 20 μl of the supernatant was loaded on 0.8 per cent agarose gel.

Genomic DNA Isolation

Genomic DNA was isolated from bacterial strains following the methods of Sambrook *et al.* (1989).

PCR Detection of Type-II PHA Synthase Genes (*phaC*1 and *phaC*2) in LDC-5 and LDC-25

The strains LDC-5 and LDC-25 yielded partial coding sequences of *phaC*1 and *phaC*2 genes. In order to identify the Type-II *pha* loci individually, primers of Solaiman (2002) were used. PCR amplification of the mixture of *phaC*1 was performed in an Eppendorf Master Cycler PCR System (Eppendorf, Hamburg, Germany) using the primer pair I-179L/DEV15R (~1.3 kb) and *phaC*2 with DEV15L/I-179R (~1.5 kb) primers (the sequences of DEV15R and DEV15L were based on consensus regions derived from a multiple sequence alignment analysis of several *Pseudomonas pha* genes). The 25 μl PCR mixture consisted of 30-50 ng of the genomic DNA template solution, 10X PCR buffer, 50 ng of each of two primers, 0.25 mM concentrations each of dATP, dCTP, dGTP and dTTP and 2.5 U of *Taq* polymerase.

Semi-nested Medium Chain Length PCR

Semi-nested MCL-PCR was performed in an Eppendorf PCR System as described by Solaiman (2002). Primers (I-179L and I-179R) were designed based on two highly conserved sequences of all Pseudomonad *pha*C genes. The primer I-179L is also homologous to a sequence region in the type I *phbC* genes, whereas, I-179R is highly specific only to the type II *phaC* genes. Purified genomic DNA or cell lysate was used as template for the PCR reaction.

Detection of PCR Products

PCR amplification mixture (10 µl) and the molecular weight marker (Gene Ruler 100 bp DNA ladder–Qiagen) were subjected to 1.0 per cent agarose gel electrophoresis (Sigma, USA) and ethidium bromide staining. The amplified DNA fragments were visualized by UV illumination and filed using a Gel-Doc photodocumentor device (Geneline, Spectronics, India).

Purification of PCR Products

PCR products were purified using the QIAquick (Qiagen, Valencia, USA) PCR purification kit according to the manufacturer's protocol. Purified DNA was recovered in 30µl of EB buffer (10 mM Tris-Cl [pH 8.5]) supplied in the kit.

Phenotypic and Biochemical Characterization

Morphological analysis, Gram and spore staining, catalase and oxidase activities, oxidation-fermentation tests, motility, starch hydrolysis, arginine utilization and Tween 80 tests were completed according to standard methods given in the *Bergey's Manual for Systematic Bacteriology* (Holt *et al.,* 1994). Cultures were also streaked onto Acetamide (selective media for *Pseudomonas*), Garibaldi's and Pyoverdin or fluorescein (greenish-yellow) medium (Luisetti *et al.,* 1972) to detect pigment production and *Pseudomonas* agar (Medium A) to detect phenazine production. Plates were incubated at 37ºC for 1-2 days. A number of additional standards such as Scanning Electron Microscopy, resistance to antibiotics, 16S rRNA sequencing and sequence analysis have been used to classify the selected isolates.

Scanning Electron Microscopy (SEM)

The bacterial cells grown under conditions favorable for PHA accumulation were washed twice by 0.5 M NaCl followed by fixing the cells in increasing concentrations of ethanol–aqueous solutions containing 30 per cent, 50 per cent, 70 per cent, 90 per cent and finally in absolute ethanol. SEM examination was performed at the Department of Plant Sciences, Madurai Kamaraj University, Madurai. The ethanol treated bacterial samples were mounted on cover glass and in turn fixed on the specimen stubs using fevicol adhesive. Small samples were mounted directly on Scotch double adhesive tape. Samples were coated with gold to a thickness of 100 Å using Hitachi Vacuum Evaporator, Model HUS 5 GB. The coated samples were analyzed in a Hitachi Scanning Electron Microscope Model S-450 (Japan) operated at 15 kV and photographed.

16S rRNA Gene Amplification and Sequencing

For 16S rRNA gene amplification, the PCR reaction mixture consisted of 30 ng of genomic DNA, 1X buffer, 3mM MgCl$_2$ 200µM each of deoxynucleotide triphosphates, 1U proof reading *Pfu* DNA polymerase (Fermentas) and 2.5µM each universal primers (16s F1: 5'-AGAGTTTGATCCTGGCTCAG-3'-*Escherichia coli* positions 8 to 27; 16sR1: 5'-ACGGCTACCTTGTTACGACTT-3'-*Escherichia coli* positions 1494 to 1513) as previously described (Weisburg *et al.,* 1991). The thermal cycle programme runs on an Eppendorf PCR System (Mastercycler Personal 5332). PCR thermal cycling was undertaken by initially denaturing DNA at 94°C for 5 min, followed by 30 cycles of 30s at 94°C, 30s at 55°C and 1 min at 72°C. A portion of the reaction mixture was used to visualize PCR products on 1 per cent (w/v) agarose gels. PCR products in the remainder of the reaction mixture were purified by QIAquick PCR purification kit (Qiagen Ltd., Crawley, West Sussex, United Kingdom) and sequenced (Direct Sequencing–Genei, Bangalore). Sequence data obtained were compared with known 16S ribosomal DNA (rDNA) sequences of *Pseudomonas* strains by using the BLAST algorithm (http: www.ncbi.nlm.nih/gov/BLAST) (Altschul *et al.,* 1990).

Phylogenetic Analysis

The 16S rRNA sequence phylogenetic data were obtained by alignment of the different 16S rRNA sequences using the program BLAST (Basic Local Alignment Search Tool) [www.ebi.ac.uk and http://www.ncbi.nlm.nih.gov], CLUSTALW (Thompson *et al.,* 1994) and BIBI (Devulder *et al.,* 2003) (http://umr5558-sud-str1.univ-lyon1.fr/lebibi/lebibi.cgi) with the default parameters. The reference sequences required for comparison were downloaded from the Genbank database based on BLAST results using the site http://www.ncbi.nlm.nih.gov/Genbank. All the sequences were aligned using multiple sequence alignment program (ClustalW) developed by Higgins *et al.,* 1992; Thompson *et al.,* 1994. Genetic relationships were determined by the distance method with the MEGA 2.1 program (Kumar *et al.,* 2001) using nucleotide sequences of the 16S rDNA gene. Distance trees were estimated according to the neighbour joining method of Saitou and Nei (1987). To determine the degree of statistical support for branches in phylogeny, 100 bootstrap replicates of data were analyzed to check the tree topology for robustness.

RNA Secondary Structure Prediction

M-fold package was used to predict RNA secondary structure (http://bioweb.pasteur.fr/seqanal/interfaces/mfold-simple.html) (Zuker, 2003).

Nucleotide Sequence Accession Number

Nucleotide sequence data determined in this study are available in the EMBL, GenBank and DDBJ databases under accession numbers: **DQ082863 (*Taxonomy ID:* 253249**) and **DQ076645 (*Taxonomy ID:* 281994**). http://www.ncbi.nlm.nih.gov/entrez/viewer.fcgi?db=nucleotide and val=91177052

Polymer Isolation from Lyophilized Bacterial Cells

Intracellular PHA polymers were extracted from lyophilized cells using the method of Mahishi *et al.,* 2003 with slight modifications as described by Sujatha and Shenbagarathai, 2006. The acetone treated freeze-dried cell mass (2.0g) was stirred with a dispersion of chloroform (25ml) and 30 per cent sodium hypochlorite (25ml) at 37°C for 90 min. The dispersion was centrifuged at 8,000 g for 20min at 30°C. Of the three phases obtained, the bottom chloroform phase containing PHA was recovered by non-solvent precipitation (ice-cold methanol, 4-6 volumes) and the recovered white powdery polymer was dissolved in chloroform for further studies.

FT-IR Spectral Acquisition

IR spectra were recorded with a SHIMADZU model 8400 S Fourier Transform Spectrophotometer (University Scientific and Instrumentation Centre (USIC), Madurai Kamaraj University, Madurai) and the spectral values collected at wave number values between 4700–250 cm^{-1} as described by Lutke-Eversloh *et al.,* 2001.

Results

PCR Amplification of Partial MCL-PHA Synthase Gene (*phaC*1 and *phaC*2)

Generally, the Polymerase Chain Reaction (PCR) provides a rapid detection of specific genes in organisms. In this study, the use of primer pair I-179L/DEV 15R and DEV15L/I-179R in the first round PCR led to the amplification of about 1.3 kb and 1.5 kb DNA fragments containing the 3' end of *phaC*1 (consists of *phaC*1 and *phaZ*) and the 5' end of *phaC*2 (*phaZ* and *phaC*2) respectively, especially when the strains LDC-5 and LDC-25 were used (Figure 7.1). A second round, semi-nested PCR was next performed on the products of the first round amplification reaction in order to obtain the specific sub-genomic sequences of *phaC*1 and *phaC*2 genes. In this semi-nested round, the primer pair I-179R/I-179L was used regardless of which primer pair was employed in the first round reaction. The semi-nested PCR reaction specifically yielded the 0.54 kb sub-genomic fragments of *phaC*1 and *phaC*2 from the first round PCR products of LDC-5 and LDC-25. These results revealed that strains LDC-5 and LDC-25 contained the type II *pha* locus.

Morphological and Biochemical Analysis of Strain LDC-5 and LDC-25

Standard microbial methods were used for the phenotypic characterization of the newly isolated strains LDC-5 and LDC-25. Biochemical analysis indicated that both LDC-5 and LDC-25 were consistent with the description of typical *Pseudomonads* according to *Bergey's Manual for Systematic Bacteriology* (Holt *et al.,* 1994), which describes the genus as being Gram-negative, straight or slightly curved rods, non-spore forming, non-mucoid, motile, catalase and oxidase positive due to the enzyme indophenol oxidase. Strain LDC-5 was isolated from field soil sample and strain LDC-25 from tannery effluent (sample group–G) subsequently incubated on LB + 2 per cent glucose agar (pH 7.0–7.2) over night at 37°C. Colonies of the strain appeared white, regular with entire margin (diameter 0.1 mm). Phase Contrast Microscopy allowed observation of Gram-negative rods (Figure 7.2a). The rods were 0.2–0.4 µm wide and 2-5µm long and showed small granules present inside the cells, a frequent feature found in *Pseudomonas*. Scanning Electron Microscopy revealed the spherical nature of the PHA granules (Figure 7.2b).

Other classical positive results shared by all the *Pseudomonas* isolates, including the reference strain, *Pseudomonas corrugata* 388 included oxidase positive reactions due to the presence of cytochrome c-oxidase in the electron transport chain, thus producing indophenol oxidase. This is one of the distinguishing characteristics in the identification of *Pseudomonas* sp. Selective plating on Acetamide (Acetyltrimethylammonium bromide) agar was designed specifically to identify

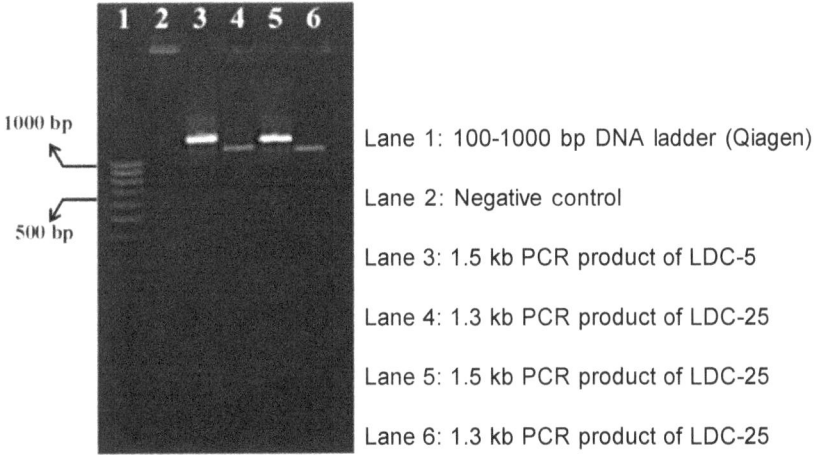

Lane 1: 100-1000 bp DNA ladder (Qiagen)

Lane 2: Negative control

Lane 3: 1.5 kb PCR product of LDC-5

Lane 4: 1.3 kb PCR product of LDC-25

Lane 5: 1.5 kb PCR product of LDC-25

Lane 6: 1.3 kb PCR product of LDC-25

Figure 7.1: Agarose gel electrophoresis of *phaC1* and *phaC2* products of strains LDC-5 and LDC-25.

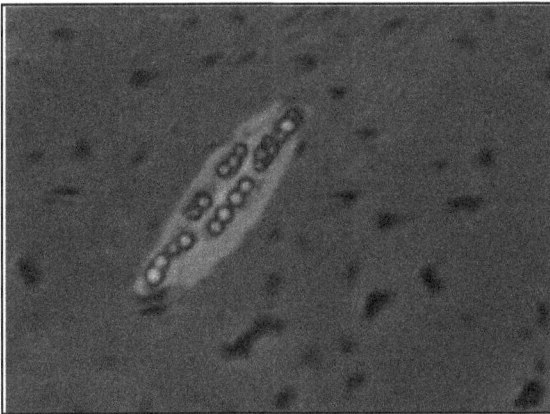

Figure 7.2a: Phase contrast microscopic view of LDC-5 with accumulated PHA granules.

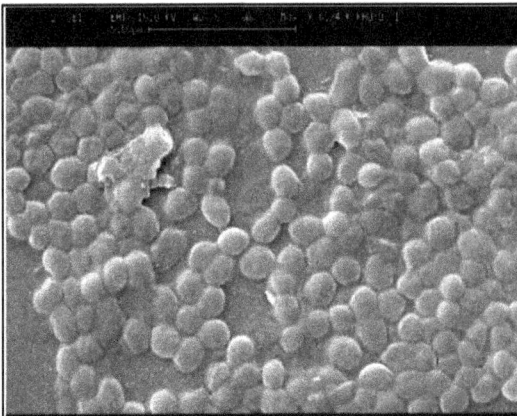

Figure 7.2b: Scanning electron microscopic view of PHA granules.

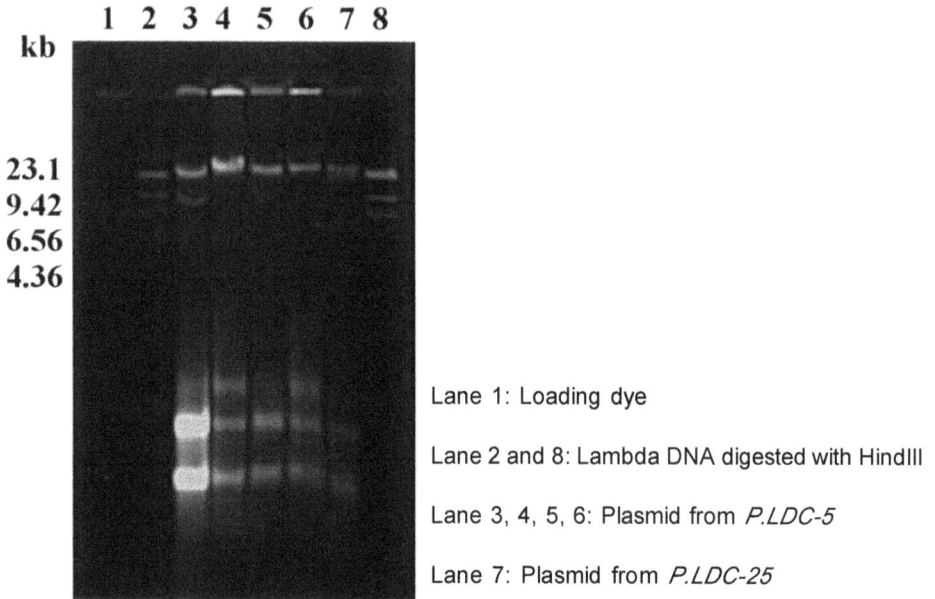

Lane 1: Loading dye

Lane 2 and 8: Lambda DNA digested with HindIII

Lane 3, 4, 5, 6: Plasmid from *P.LDC-5*

Lane 7: Plasmid from *P.LDC-25*

Figure 7.3: Plasmid profile of LDC-5 and LDC-25.

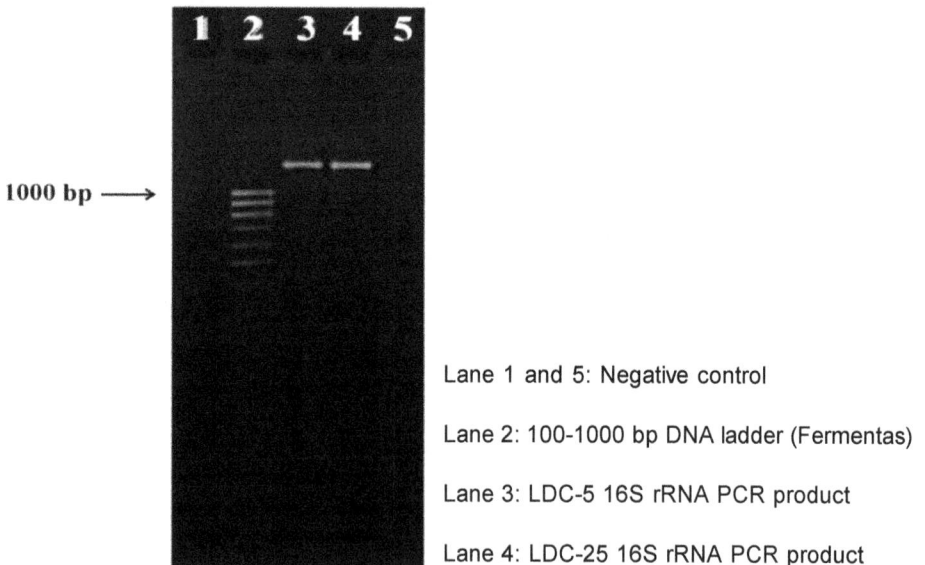

Lane 1 and 5: Negative control

Lane 2: 100-1000 bp DNA ladder (Fermentas)

Lane 3: LDC-5 16S rRNA PCR product

Lane 4: LDC-25 16S rRNA PCR product

Figure 7.4: Agarose gel electrophoresis of 16S rRNA PCR product of LDC-5 and LDC-25.

Pseudomonas species and to prevent growth of other organisms; and with addition of nalidixic acid to the selective medium, LDC-5 and LDC-25 were reconfirmed as *Pseudomonas*. Approximately 90 isolates (21 per cent) showed obvious fluorescence, but LDC-5 and LDC-25 did not and the colony characteristics of LDC-5 and LDC-25 strains were identical. Both the strains gave positive results for arginine hydrolysis, growth on Acetamide agar and non-fluorescence on *Pseudomonas agar* (Medium A) and possessed lipolytic activity (Tween 80 hydrolysis), and were not capable of hydrolysing starch. Furthermore, they were found to be very active in degradation of casein, indicating the presence of proteases among its hydrolytic enzymes. The strains were negative for the tests of methyl red, starch, hydrogen production from H_2S and growth on MacConkey medium. Carbohydrate dehydration at 30°C for 48 h was positive for D-glucose and sucrose. These tests were selected as conventional phenotypic tests used to identify *Pseudomonas* organisms (Costas, 1992). Under conditions of excess carbon (1 per cent), highly refractile intracellular poly beta granules could be observed by Phase Contrast and Fluorescence Microscopy. The spectrum of usable carbon sources were determined by various experimental procedures. Strains LDC-5 and LDC-25 were unable to grow on Minimal Media supplemented with Arabinose, Rhamnose, Salicin and Galactose. Susceptibility of the isolates was assessed by resistance to different antibiotics (Ampicillin, Kanamycin, Tetracycline, Nalidixic acid, Chloramphenicol, Gentamycin, Penicillin and Rifampcin). The results obtained through the morphological and biochemical tests assigned the strain LDC-5 and LDC-25 to the phenotypic group of the genus *Pseudomonas*. Even then, as previously stated by other authors, identification of any microbe to its species level is

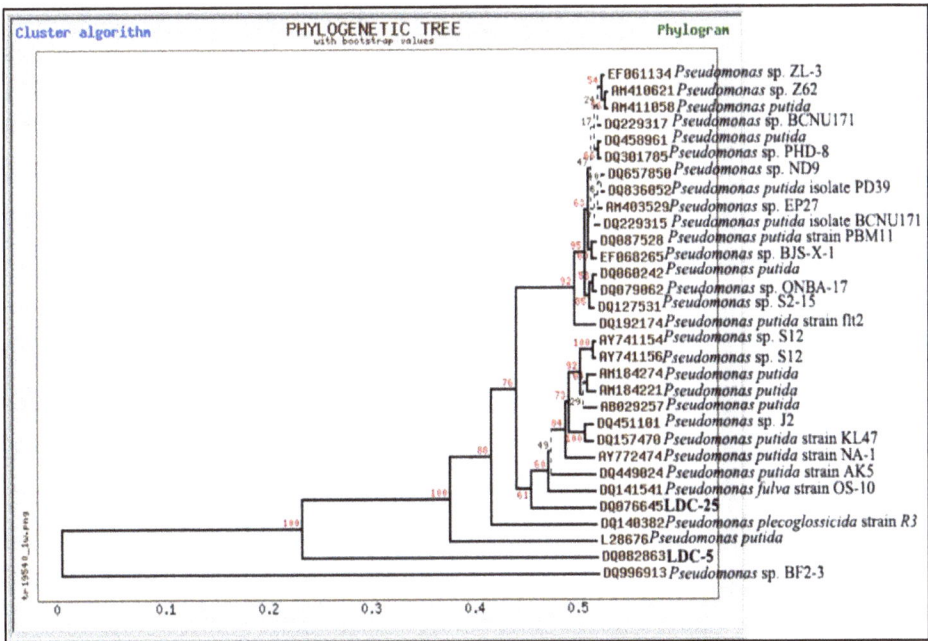

Figure 7.5. Phylogenetic tree constructed using neighbor-joining algorithm based on LDC-5 and LDC-25 16S rRNA sequence.

Numbers adjacent to branch points are bootstrap percentages (n=100 replicates). Analysis was performed by including other 16S rRNA sequences deposited in GenBank (Organisms details are indicated in parentheses). Bar=0.05 estimated substitution per sequence position.

difficult by traditionally used methods (Goodfellow, 1989). For this reason, the biochemical tests were followed by the most recent powerful 16S rRNA sequence analysis for further taxonomic confirmation.

Plasmid Profile of *Pseudomonas* sp. LDC-5 and LDC-25

Most plasmids are small, circular, supercoiled and about 0.2 to 4 per cent of the size of the bacterial chromosome. Figure 7.3 shows the results of agarose gel electrophoresis analysis of the plasmid DNA from strain *Pseudomonas* sp. LDC-5 and LDC-25. The plasmid profile of the bacteria exhibited three major bands (23 kb, 8 kb and 3 kb) and three minor bands (2 kb and below 2 kb) while running parallel to the λ DNA digested with Hind III marker. The resistance to antibiotic may be due to the presence of this plasmid, which in turn is encoded by the antibiotic resistance genes.

16S rRNA Sequence Analysis

Two strains such as LDC-5 and LDC-25 that possessed the Type II *pha* operon were further characterized by a comparative 16S rRNA gene sequence analysis. Almost entire sequences of this gene LDC-5 (DQ082863 -1,602 bp) and LDC-25 (DQ076645– ~1436 bp) were obtained (Figure 7.4) and compared with known 16S rRNA sequences published in the Ribosomal database. The neighbour-joining phylogenetic tree demonstrated that strain LDC-5 had high percentage of identity (97 per cent) with the type strains of *Pseudomonas putida* (L28676), *Pseudomonas* sp. BF2-3 (DQ996913) and *Pseudomonas plecoglossicida* strain R3 (DQ140382) (Table 7.1 and Figure 7.5) and strain LDC-25 was positioned with *Pseudomonas fulva* (DQ141541 -97 per cent) (Figure 7.5) probably indicating that both these species are taxonomically related to *Pseudomonas putida*.

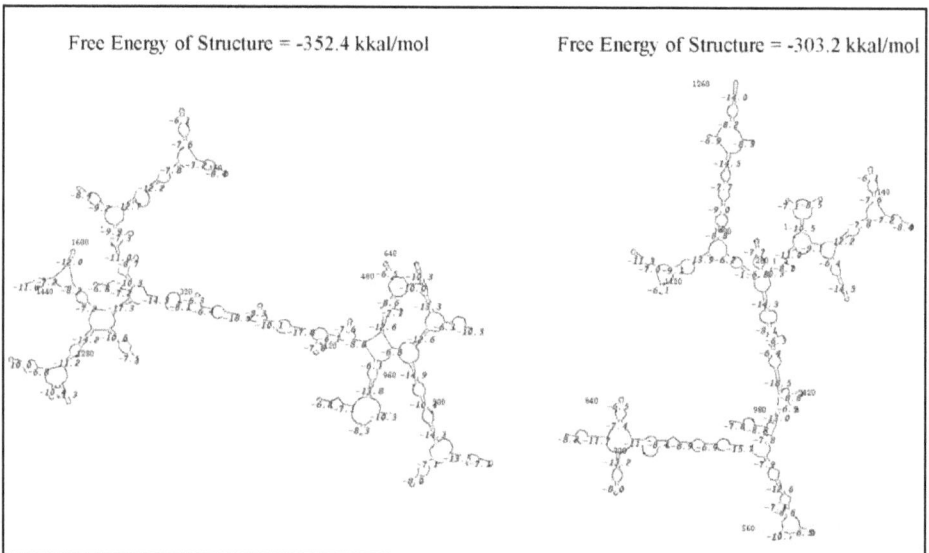

Figure 7.6: LDC-5 and LDC-25 16S rRNA secondary structure prediction using Mfold tool.

Table 7.1: Nucleotide sequences used to generate phylogenetic tree for LDC-5 and LDC-25 16S rRNA gene sequences.

Sl.No.	Accession Number	Organisms Detail	Length (bp)	Identity (per cent)
1.	DQ082863	LDC-5	1602	100
2.	DQ076645	LDC-25	1436	98
3.	DQ157470	*Pseudomonas putida* strain KL47	1456	97
4.	DQ140382	*Pseudomonas plecoglossicida* strain R3	1403	97
5.	EF068265	*Pseudomonas* sp. BJS-X-1	1495	97
6.	EF061134	*Pseudomonas* sp. ZL-3	1501	97
7.	DQ996913	*Pseudomonas* sp. BF2-3	1102	97
8.	DQ836052	*Pseudomonas putida* isolate PD39	1499	97
9.	DQ657850	*Pseudomonas* sp. ND9	1499	97
10.	DQ458961	*Pseudomonas putida*	1501	97
11.	DQ451101	*Pseudomonas* sp. J2	1456	97
12.	DQ301785	*Pseudomonas* sp. PHD-8	1500	97
13.	DQ229316	*Pseudomonas* sp. BCNU171	1499	97
14.	DQ229315	*Pseudomonas putida* isolate BCNU106	1499	97
15.	DQ192174	*Pseudomonas putida* strain flt2	1515	97
16.	DQ133572	*Pseudomonas* sp. ND24	1491	97
17.	DQ127531	*Pseudomonas* sp. s2-15	1500	97
18.	DQ127530	*Pseudomonas* sp. J19	1495	97
19.	DQ091247	*Pseudomonas* sp. FL40	1080	97
20.	DQ087528	*Pseudomonas putida* strain PBM11	1498	97
21.	DQ079062	*Pseudomonas* sp. ONBA-17	1501	97
22.	DQ060242	*Pseudomonas putida*	1502	97
23.	AY772474	*Pseudomonas putida* strain NA-1	1463	97
24.	AY741156	*Pseudomonas* sp. S12	1480	97
25.	AY741154	*Pseudomonas* sp. S12	1480	97
26.	AY647158	*Pseudomonas putida* strain ZWL73	1423	97
27.	AM411070	*Pseudomonas* sp. Z64-1	1501	97
28.	AM411058	*Pseudomonas putida*	1501	97
29.	AM410621	*Pseudomonas* sp. Z62	1501	97
30.	AM403529	*Pseudomonas* sp. EP27	1499	97
31.	AM184274	*Pseudomonas putida*	1474	97
32.	AM184239	*Pseudomonas putida*	1472	97
33.	AM184221	*Pseudomonas putida*	1483	97
34.	AF378011	*Pseudomonas* sp. ML2	1501	97

Contd...

Table 7.1–*Contd...*

Sl.No.	Accession Number	Organisms Detail	Length (bp)	Identity (per cent)
35.	DQ481473	*Pseudomonas putida* strain VTs-23	1425	97
36.	DQ141541	*Pseudomonas fulva* strain OS-10	1472	97
37.	DQ229317	*Pseudomonas putida* isolate BCNU154	1500	97
38.	DQ449024	*Pseudomonas putida* strain AK5	1428	97
39.	AB029257	*Pseudomonas putida*	1470	97
40.	L28676	*Pseudomonas putida*	1330	97
41.	EF051575	*Pseudomonas putida* strain T7-5	1550	97

The present study provided a secondary structure model for 16S ribosomal RNA of *Pseudomonas* sp. LDC-5 (1602 residues) and *Pseudomonas* sp. LDC-25(1436 residues) on the basis of comparative sequence analysis. It is interesting to note the presence of potential helical structure in LDC-5 at the following regions (Free energy is indicated in brackets): 616-621/682-677 (-13.300000 Kkal/mol); 786-791/857-852 (-12.600000 Kkal/mol); 948-951/959-956 (-7.800000 Kkal/mol); 1240-1243/1255-1252 (-7.500000 Kkal/mol) and 1422-1424/1483-1481 (-8.300000 Kkal/mol). Comparatively LDC-25 showed the presence of potential helices at 633-637/911-907 (-6.900000 Kkal/mol); 1219-1221/1241-1239 (-6.900000 Kkal/mol); 843-845/875-873 (-6.500000 Kkal/mol); 957-960/968-965 (-7.800000 Kkal/mol) (Figure 7.6).

FT-IR Spectral Analysis

Fourier Transform Infra red Spectroscopy is a sensitive and non-destructive workhorse technique. The IR spectrum of a sample represents its total chemical

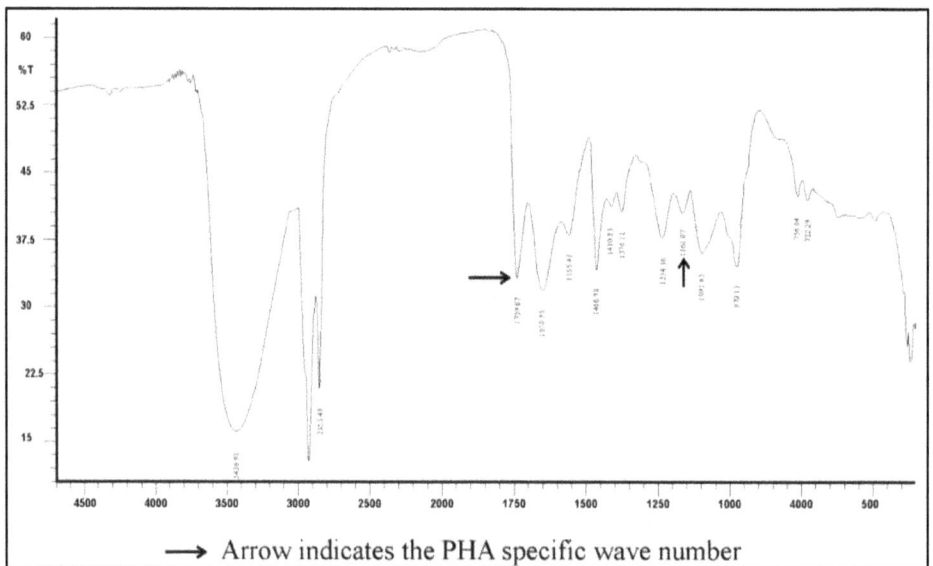

→ Arrow indicates the PHA specific wave number

Figure 7.7: FT-IR spectral analysis of PHA recovered from LDC-5.

composition, because every chemical compound in the sample makes its own distinct contribution to the absorbance spectrum. The distinctness of an individual spectrum which is determined by the chemical structure of each component and the degree, to which each component contributed to the spectrum, is directly related to the concentrations of the components of the sample. The representative spectra in Figure 7.7 are the results of the cumulative absorbance of LDC-5 and LDC-25 cells. These spectra are dominated by the absorbance of the major cellular constituent, namely PHA. The major PHA bands are the intense ester carbonyl stretch at 1738 to 1728 cm^{-1}, a number of strong bands at wave number values between 1,450 and 1,000 cm^{-1} are due to methyl (CH_3) and methylene (CH_2) deformations and C-O stretches. Although other esters, such as wax esters or triacylglycerols, may also generate similar peaks, no such disturbance was observed when cells of *E.coli* were exposed to FTIR. This further demonstrated that the band between 1738 to 1728 cm^{-1} is characteristic of PHA.

Discussion

In order to overcome the high production costs of biologically produced PHA, a strategy for identifying and isolating suitable PHAs producing microorganisms from the environment is therefore considered a most critical factor in the PHA production process (Kung *et al.,* 2007). Recent applications of molecular biology and molecular genetic techniques, such as total DNA isolation and characterization, G + C composition, PCR amplification of rRNA sequences and functional genes, *in situ* hybridization of rRNA oligonucleotide probes have provided tools to determine microbial presence and diversity in the environment (Atlas *et al.,* 1992; Akkermans *et al.,* 1995; Head *et al.,* 1998). Comparison of nucleotide sequences have shown that there are regions of rRNA sequences that are highly conserved among all organisms and other regions that vary to different degrees. The variability in these regions increases as the evolutionary distance between two organisms increases, which provide a means to determine phylogenetic relationships and to distinguish microorganisms from one another (Woese, 1987; Woese, 1992). Presently, the rRNA genes in DNA taken directly from soil can be amplified using PCR, the products cloned, and the nucleotide sequence determined. By employing these techniques, many scientists have begun to examine the biodiversity of soil microbial communities (Ueda *et al.,* 1995; Borneman and Triplett, 1997; Jurgens *et al.,* 1997).

Systematic and analytical techniques for identification have evolved due to advancements in molecular techniques. These techniques particularly, when combined with conventional analyses have enabled rapid and definitive identification of unknown isolates from soil community. Therefore, in this study, conventional identification methods followed by 16S rRNA gene sequence analysis were employed to identify the taxonomic position of the selected strains. The two strains LDC-5 and LDC-25 isolated from field and effluent samples respectively possessed Type–II PHA synthase genes as confirmed by gene amplification using Medium Chain Length specific primer, I-179L/DEV15R and DEV15L/I-179R (Figure 7.1) (Solaiman, 2002). A second round semi-nested PCR yielded 0.54 kb specific subgenomic fragments of *phaC1* and *phaC2* from the first round MCL-PCR products obtained, further supported the observation that MCL-PHAs are mostly produced by *Pseudomonas* belonging to the rRNA-DNA homology group I. These results are in accordance with the previous works of Steinbuchel, 1991; Solaiman *et al.,* 2000; Rehm, 2003; 2007. In addition, the earlier reports of Gerngross *et al.,* 1994, Timm and Steinbuchel, 1992, revealed that these subgenomic fragments contain information surrounding the catalytically important lipase box like sequence and the phosphopantethine binding site. Preliminary morphological, biochemical and 16S rRNA analysis also positioned the strains in Pseudomonads family.

Although the conventional (Figures 7.2a-b, 7.3) and comprehensive automated biochemical tests (carried out at Microbial Type Culture Collection and Gene bank (MTCC)) preliminarily identified the isolates to be of the *Pseudononas* group, both systems have innate problems that can especially, lead to misidentification of strains. To circumvent misidentification issues, nowadays, identification and classification of bacteria from soil samples at the species and higher taxonomic levels mainly relies on a genotypic approach, typically involving analysis of ribosomal RNA gene (rRNA). The backbone of the current taxonomy of prokaryotes is almost exclusively based upon a phylogenetic network derived from comparative sequence analysis of the small subunit rRNAs and respective phylogenetic marker genes (Ludwig and Klenk, 2001). As 'living fossils', these molecules roughly reflect the evolutionary history of the respective organisms (Gee *et al.,* 2003; Wellinghausen *et al.,* 2005; Kumar *et al.,* 2001). 16s rRNA sequencing is therefore now considered the most precise method for phenotypic identification (Woese, 1987). It was for this reason that the 16S rRNA sequencing was performed on LDC-5 and LDC-25 isolates for definitive identification (Figure 7.4). Though 16S rRNA genes are conserved among all organisms they possess various unique species regions that allow for bacterial identification (Gobel *et al.,* 1987). However, sequencing the entire sequence is desirable and usually required when describing a new species (Clarridge, 2004). In the present study, nearly complete sequences of the 16S rRNA gene of the selected strains were determined and compared with those available from Ribosomal data base. Results of this comparison revealed that strain LDC-5 was most similar to *Pseudomonas putdia* and *Pseudomonas plecoglossicida* strain R3 (subclass of proteobacteria) (Table 7.1 and Figure 7.5), whereas, strain LDC-25 was closely related to *Pseudomonas fulva* (subclass of proteobacteria) (Figure 7.5). Traditionally, according to Palleroni (1984), *Pseudomonas* species were classified into one of five natural groups based on RNA homology, the so-called 'RNA homology groups'. Group I composed of fluorescent as well as non-fluorescent species that are also known as *Pseudomonas* sensu stricto, and the other groups consisted of various, often pathogenic species. Recently, along with progress in molecular techniques, the classification of Pseudomonads has been revised and members of group I are retained in the genus *Pseudomonas*, but members of groups II, III, IV and V are moved to new genera (Moore *et al.,* 1996). Based on 16S sequencing data and on biochemical criteria (Grimont *et al.,* 1996), Pseudomonads are divided into two groups whose representative species are *Pseudomonas aeruginosa* and *Pseudomonas fluorescens*. The strains LDC-5 and LDC-25 characterized in this study fall into group I (RNA homology group) Pseudomonads.

Ribosomal RNA (rRNA), a key molecule in protein synthesis with its ubiquitous presence and highly conserved nature, is the basis for modern molecular taxonomy.

Ribosomal RNA function is largely determined by its structure (Noller, 1981) as the general structure of rRNA is universally conserved across all taxa. The secondary structure of rRNA, even when not universally identical across taxa, is more highly conserved than are nucleotides (Woese, 1987; Gutell *et al.,* 1985). It is interesting to note that helices at 800-852 regions are predicted to be much more stable and regions of 946-955/1225-1235 is involved in formation of long range helix and single base bulge. The helices involving positions 600, 830 and 1240 are involved in large type of helix. The highly conserved regions; 783-786/796-799 and 960-963/972-975 involved in potential helix formation. Hence, the results of this study (Figure 7.6) are in good agreement with the reports of Woese *et al.,* 1980.

Morphological, biochemical, 16S rRNA and PHA specific PCR analyses revealed that the selected strains were significant in terms of their Polyhydroxyalkanoate production. In this regard, for further confirmation, qualitative analysis was carried out using the recovered polymer from LDC-5 and LDC-25. As per review, there are several methods that have been developed for qualitative analysis of PHA, including: GC, NMR and pyrolysis (Braunegg *et al.,* 1978; Cross *et al.,* 1989). These methods often require extensive and complicated sample preparations like hydrolysis, extraction, purification or methylation etc. On the other hand, FT-IR technique has several advantages including no solvent requirement, minimal simple sample preparation and greatly reduced analysis time (~ 30 min). FT-IR technique also presents a rapid screening tool for new organisms and their abilities to synthesize various types of PHA (Kansiz *et al.,* 2000). The recovered PHA from LDC-5 and LDC-25 showed their strongest ester carbonyl band at 1735 cm^{-1} to 1744 cm^{-1} (Figure 7.7) supporting the earlier studies of Hong *et al.,* 1999. Bands at 970, 1234 cm^{-1} are characteristic of the crystalline phase of the copolymer and the bands at 1739 cm^{-1} are characteristic of the amorphous phase. These results are in good agreement with the report of Luo *et al.,* 2006. In addition, from the FT-IR spectrum, it has been realized that the peaks at 1739.67, 1161.07, 2853.49 cm^{-1} revealed the possibility of HB and MCL (HA) components in the polymer as suggested by Hong *et al.* (1999).

Conclusion

The results of the present study demonstrated that the taxonomic characterization of selected PHA accumulators (LDC-5 and LDC-25) by conventional biochemical and 16S rRNA sequence analysis. The incorporation of conventional biochemical identification methods into the isolation strategy proves to be a useful approach in determining the isolates in their genus level. The strain LDC-5 is positioned with the type strain of *Pseudomonas putida* (L28676-97 per cent), whereas LDC-25 phylogenetically is closer to *Pseudomonas fulva* (DQ141541-97 per cent) strain. 16S rRNA secondary structure prediction revealed the presence of potential helical structure in LDC-5 and LDC-25. Fourier transform infrared (FT-IR) spectroscopy revealed that the recovered PHA belongs to SCL-MCL-PHA. This opens up new possibilities for various industrial applications owing to the superior properties of this new co-polymer.

Acknowledgements

The authors are thankful to Department of Science and Technology (DST-WOS-A No. SR/WOS-A/LS-137/2004), New Delhi for financial support.

References

Akkermans, A., van Elsas, J.D. and de Bruijn, F.J. (Eds.), 1995. *Molecular Microbial Ecology Manual.* Kluwer Academic Publ., Nowell, MA.

Altschul, S.F., Gish, W., Miller, W., Myers, E. W. and Lipman, D.J., 1990. Basic local alignment search tool. *J. Mol. Biol.*, 215: 403–410.

Anderson, A.J. and Dawes, E.A., 1990. Occurrence, metabolism, metabolic role, and industrial uses of bacterial polyhydroxyalkanoates. *Microbiol Rev.*, 54: 450–472.

Ashby, R.D., Solaiman, D.K.Y. and Foglia, T.A., 2002. The synthesis of short- and -medium-chain length poly (hydroxyalkanoate) mixtures from glucose or alkanoic acid-grown *Pseudomonas oleovorans. J. Ind. Microbiol. Biotechnol.*, 28: 147–153.

Atlas, R.M., 1993. *Microbial Ecology: Fundamentals and Applications,* 3rd Edn. p. 39–43.

Atlas, R.M., Sayler, G.S., Burlage, R.S. and Bej, A.K., 1992. Molecular approaches for environmental monitoring of microorganisms. *Bio Techniques*, 12: 706–717.

Baker, G.C., Beebee, T.J.C. and Ragan, M.A., 1999. *Prototheca richardsi:* A pathogen of anuran larvae, is related to a clade of protistan parasites near the animal-fungal divergence. *Microbiology,* : 1777–1784.

Borneman, J. and Triplett, E.W., 1997. Molecular microbial diversity in soils from eastern Amazonia—evidence for unusual microorganisms and microbial population shifts associated with deforestation. *Appl. Environ. Microbiol.*, 63: 2647–2653.

Bottger, E. C., 1989. Rapid determination of bacterial ribosomal RNA sequences by direct sequencing of enzymatically amplified DNA. *FEMS Microbiol. Lett.*, 65: 171–176.

Braunegg, G., Sonnleitner, B., Lafferty, R.M., 1978. A rapid gas chromatographic method for the determination of poly-β-hydroxybutyrate in microbial biomass. *Eur. J. Appl. Microbiol.*, 6: 29–37.

Chen, J.Y., Song, G. and Chen, G.Q., 2006. A lower specificity PhaC2 synthase from *Pseudomonas stutzeri* catalyses the production of co-polyesters consisting of short-chain-length and medium-chain-length 3-hydroxyalkanoates. *Antonie Van Leeuwenhoek*, 89: 157–167.

Clarridge, J.E. (III), 2004. Impact of 16S rRNA gene sequence analysis for identification of bacteria on clinical Microbiology and Infectious Diseases. *Clinical Microbiol. Rev.*, 17: 840–862.

Costas, M., 1992. Classification, identification and typing of bacteria by the analysis of their one-dimensional polyacrilamide gel electrophoretic protein patterns. *Advanced Electrophoresis,* 5: 351–408.

Cross, R.A., DeMello, C., Lenz, R.W., Brandl, H. and Fuller, R.C., 1989. Biosynthesis and characterization of poly (β-hydroxyalkanoates) produced by *Pseudomonas oleovorans. Macromolecules*, 22: 106–1115.

Devereux, J.P., Haeberli, P. and Smithies, O., 1984. A comprehensive set of analysis programs for the VAX. *Nucl Acid Res.*, 12: 387–395.

Doi, Y., 1990. *Microbial Polyesters*. VCH Publishers, N.Y.

Garrity, G.M. and Holt, J.G., 2001. The road map to the manual. In: *Bergey's Manual of Systematic Bacteriology*, (Ed.) G.M. Garrity. Springer-Verlag, New York, p. 119–166.

Gee, J.E., Sacchi, C.T., Glass, M.B., De, B.K., Weyant, R.S., Levett, P.N., Whitney, A.M., Hoffmaster, A.R. and Popovic, T., 2003. Use of 16S rRNA gene sequencing for rapid identification and differentiation of *Burkholderia pseudomallei* and *B. mallei. J. Clin. Microbiol.*, 41: 4647–4654.

Gerngross, U.T., Snell, K.D., Peoples, O.P., Sinskey, A.J., Csuhai, E., Masamune, S. and Stubbe, J., 1994. Overexpression and purification of the soluble Polyhydroxyalkanoate synthase from *Alcaligenes eutrophus:* Evidence for a required posttranslational modification for catalytic activity. *Biochemistry,* 33: 9311–9320.

Gobel, U.B., Geiser, A. and Stanbridge, E.J., 1987. Oligonucleotide probes complementary to variable regions of ribosomal RNA discriminate between Mycoplasma species. *J. Gen. Microbiol.,* 133: 1969–1971.

Goodfellow, M., 1989. Genus *Rhodococcus*. In: *Bergey's Manual of Systematic Bacteriology*, Vol. 4, (Eds.) S.T. Williams, M.E. Sharpe and J.G. Holt. Williams and Wilkins, Baltimore, pp. 2362–2371.

Grimont, P.A.D., Vancanneyt, M., Lefevre, M., Vandemeulebroecke, K., Vauterin, L., Brosch, R., Kersters, K. and Grimont, F., 1996. Ability of biolog and biotype-100 systems to reveal the taxonomic diversity of the Pseudomonads. *Syst. Appl. Microbiol.*, 19: 510–527.

Harmsen, D. and Karch, H., 2004. 16S rDNA for diagnosing pathogens: A living tree. *ASM News,* 70: 19–24.

Head, I.M., Saunders, J.R. and Pickup, R.W., 1998. Microbial evolution, diversity, and ecology: A decade of ribosomal RNA analysis of uncultivated microorganisms. *Microbial. Ecol.*, 35: 1–21.

Higgins, D.G., Bleasy, A.J. and Fuchs, A.J., 1992. CLUSTALW: Improved software for multiple sequence alignment. *Comput. Appl. Biosci.,* 8: 189–191.

Holt, J.G., Krieg, N.R., Sneath, P.H.A., Staley, J.T. and Williams, S.T., 1994. *Bergey's Manual of Determinative Bacteriology*, 9th Edn. Williams and Wilkins, Baltimore.

Hong, K., Sun, S., Tian, W., Chen, G.Q. and Huang, W., 1999. A rapid method for detecting bacterial Polyhydroxyalkanoates in intact cells by Fourier Transform Infrared Spectroscopy. *Appl. Microbiol. Biotechnol.,* 51: 523–526.

Hugenholtz, P., Goebel, B.M. and Pace, N.R., 1998. Impact of culture-independent studies on the emerging phylogenetic view of bacterial diversity. *J. Bacteriol.*, 180: 4765–4774.

Jensen, G.B., Wilcks, A., Petersen, S.S., Damgaard, J., Baum, J.A. and Andrup, L., 1995. The genetic basis of the aggregation system in *Bacillus thuringiensis* subsp. *israelensis*. Host range and kinetics. *Curr. Microbiol.*, 33: 228–236.

Jurgens, G., Lindstrom, K. and Saano, A., 1997. Novel group within the kingdom Crenarchaeota from boreal forest soil. *Appl. Environ. Microbiol.*, 63: 803–805.

Kansiz, M., Billman-Jacobe, H. and McNaughton, D., 2000. Quantitative determination of the biodegradable polymer poly (β–hydroxybutyrate) in a recombinant *Escherichia coli* strain by use of mid-infrared spectroscopy and multivariate statistics. *Appl. Environ. Microbiol.*, 66: 3415–3420.

Kumar, S., Tamura, K., Jakobsen, I.B. and Nei, M., 2001. MEGA2: Molecular evolutionary genetics analysis ware. *Bioinformatics*, 17: 1244–1245.

Kung, S.S., Chuang, Y.C., Chen, C.H. and Chien, C.C., 2007. Isolation of polyhydroxyalkanoates-producing bacteria using a combination of phenotypic and genotypic approach. *Lett. Appl. Microbiology*, 44: 364–371.

Ludwig, W. and Klenk, H.P., 2001. Overview: aAphylogenetic backbone and taxonomic framework for prokaryotic systematics. In: *Bergey's Manual of Systematic Bacteriology*, (Ed.) G. Garrity. Springer, New York, pp. 49–65.

Luo, R., Chen, J., Zhang, L. and Chen, G.Q., 2006. Polyhydroxyalkanoates co polyesters produced by *Ralstonia eutropha* PHB^{-4} harboring a low-substrate specificity PHA synthase PhaC2Ps from *Pseudomonas stutzeri* 1317. *Biochem. Eng. J.*, 32: 218–225.

Lutke-Eversloh, T., Bergander, K., Luftmann, H. and Steinbuhel, A., 2001. Identification of a new class of biopolymer: Bacterial synthesis of a sulfur-containing polymer with thioester linkages. *Microbiology*, 147: 11–19.

Matsusaki, H., Manji, S., Taguchi, K., Kato, M., Fukui, T. and Doi, Y., 1998. Cloning and molecular analysis of the poly (3-hydroxybutyrate) and poly (3-hydroxybutyrate-co-3-hydroxyalkanoate) biosynthesis genes in *Pseudomonas* sp. strain 61–3. *J. Bacteriol.*, 180: 6459–6467.

Moore, E.R.B., Mau, M., Arnscheidt, A., Bottger, E.C., Hutson, R.A., Collins, M.D., Van de Peer, Y., de Wachter, R. and Timmis, K.N., 1996. The determination and comparison of the 16s rRNA gene sequences of species of the genus *Pseudomonas* (*Sensu stricto*) and estimation of the natural intrageneric relationships. *Syst. Appl. Microbiol.*, 19: 478–492.

Noller, H.F. and Woese, C.R., 1981. Secondary structure of 16S ribosomal RNA. *Science*, 212: 403–411.

Palleroni, N.J., 1992. Introduction to the family *Pseudomonadaceae*. In: *The Prokaryotes: A Handbook on the Biology of Bacteria: Ecophysiology, Isolation, Identification, Applications*, Vol. 3, (Eds.) A. Balows, H.G. Truper, M. Dworkin, W. Harder and K.H. Schleifer. Springer, New York, pp. 3071–3085.

Rehm, B.H.A., 2003. Polyester synthases: Natural catalysts for plastics. *Biochemistry J.*, 376: 15–33.

Rehm, B.H.A., 2007. Biogenesis of microbial polyhydroxyalkanoates granules: A platform technology for the production of tailor-made bioparticles. *Curr. Issues Mol. Biol.*, 19: 41–62.

Rothermich, M.M., Guerrero, R. and Lenz, R.W., 2000. Characterization, seasonal occurrence, and diel fluctuation of poly (hydroxyalkanote) in photosynthetic microbial mats. *Appl. Environ. Microbiol.*, 10: 4279–4291.

Saitou, N. and Nei, M., 1987. Neighbour-joining method: A new method for reconstructing phylogenetic trees. *Molecular Biology and Ecology*, 4: 406–425.

Shamala, T.R., Chandrashekar, A., Vijayendra, S.V. and Kshama, L., 2003. Identification of polyhydroxyalkanoate (PHA)–producing *Bacillus* sp. using the polymerase chain reaction (PCR). *J. Appl. Microbiol.*, 94: 369–374.

Sheu, D.S., Wang, Y.T. and Lee, C.Y., 2000. Rapid detection of polyhydro-xyalkanoates accumulating bacteria isolated from the environment by colony–PCR. *Microbiology,* 146: 2019–2025.

Solaiman, D.K.Y., 2000. PCR cloning of *Pseudomonas resinovorans* polyhydro-xyalkanote biosynthesis genes and expression in *Escherichia coli. Biotechnol. Lett.*, 22: 789–794.

Solaiman, D.K.Y., 2002. Polymerase-chain-reaction-based detection of individual polyhydroxyalkanoate synthase *phaC1* and *phaC2* genes. *Biotechnol. Lett.,* 24: 245–250.

Solaiman, D.K.Y. and Ashby, R.D., 2005. Rapid genetic characterization of poly (hydroxyalkanoate) synthase and its applications. *Biomacromolecules,* 6: 532–537.

Solaiman, D.K.Y., Ashby, R.D. and Foglia, T.A., 2000. Rapid and specific identification of medium-chain-length polyhydroxyalkanoate synthase gene by polymerase chain reaction. *Appl. Microbiol. Biotechnol.*, 53: 690–694.

Spiekermann, P., Rehm, B.H.A., Kalscheuer, R., Baumeister, D. and Steinbuchel, A., 1999. A sensitive, viable colony staining method using Nile red for direct screening of bacteria that accumulate polyhydroxyalkanoic acids and other lipid storage compounds. *Arch. Microbiol.*, 171: 73–80.

Steinbuchel, A., 1991. *Biomaterials: Novel Materials from Biological Sources. Polyhydroxyalkanoic Acid,* (Ed.) D. Byrom. Stockton, New York, p. 124–213.

Steinbuchel, A. and Schlegel, H.G., 1991. Physiology and molecular genetics of poly (β-hydroxyalkanoic acid) synthesis in *Alcaligenes eutrophus. Mol. Microbiol.*, 5: 535–542.

Sujatha, K. and Shenbagarathai, R., 2006. A Study on MCL-Polyhydroxyalkanoate accumulation in *E. coli* harboring *phaC1* gene of indigenous *Pseudomonas* sp. LDC–5. *Lett. Appl. Microbiology,* 43: 607–614.

Thompson, J.D., Higgins, D.G. and Gibson, T.J., 1994. CLUSTAL W: Improving the sensitivity of progressive multiple sequence alignment through sequence weighting, positions–specific gap penalties and weight matrix choice. *Nucl. Acids Res.*, 22: 4673–4680.

Timm, A. and Steinbuchel, A., 1992. Cloning and molecular analysis of the poly-β-hydroxyalkanoic acid gene locus of *Pseudomonas aeruginosa* PAO1. *Eur. J. Biochem.*, 209: 15–30.

Ueda, T., Suga, Y. and Matsuguchi, T., 1995. Molecular phylogenetic analysis of a soil microbial community in a soybean field. *Eur. J. Soil Sci.*, 46: 415–421.

Van Beilen, J.B., Panke, S., Lucchini, S., Franchini, A.G., Rothlis-berger, M. and Witholt, B., 2001. Analysis of *Pseudomonas putida* alkane-degradation gene clusters and flanking insertion sequences: Evolution and regulation of the *alk* genes. *Microbiology,* 147: 1621–1630.

Wellinghausen, N., Kothe, J., Wirths, B., Sigge, A. and Poppert, S., 2005. Superiority of molecular techniques for identification of gram-negative, oxidase-positive rods, including morphologically nontypical *Pseudomonas aeruginosa*, from patients with cystic fibrosis. *J. Clin. Microbiol.*, 43: 4070–4075.

Woese, C., 1992. Prokaryote systematics: The evolution of the science. In: *The Prokaryotes*, (Ed.) Truper H.G., *et al.* Springer Verlag, New York, p. 3–18.

Woese C.R., 1987. Bacterial evolution. *Microbiol. Rev.*, 51: 221–271.

Woese, C.R., Magrun, L.J., Gupta, R., Siegel, R.B. and Stahl, D.A., 1980. Secondary structure model for bacterial 16S ribosomal RNA. *Phylogenetic, Enzymatic and Chemical Evidence*, 8: 2275–2293

Wu, H.A., Sheu, D.S. and Lee, C.Y., 2003. Rapid differentiation between short-chain-length and medium-chain-length polyhydroxyalkanoates-accumulating bacteria with spectrofluorometry. *J. Microbiol. Methods*, 53: 131–135.

Zuker, M., 2003. Mfold web server for nucleic acid folding and hybridization prediction. *Nucl. Acids Res.*, 31: 3406–3415.

Editor: **Dr. K. Muthuchelian,** *Vice Chancellor, Periyar University, Salem*
Published by: **Daya Publishing House, NEW DELHI**

Chapter 8

DNA Barcoding in Plants: An Advanced Technology for Biodiversity Inventory

P.N. Krishnan

Biotechnology and Bioinformatics Division,
Tropical Botanic Garden and Research Institute, Trivandrum

Biodiversity represent the totality of inherited variation of all life forms of life (Plants, Animals and Microbes) across all known hierarchial levels: genes, species and ecosystems. Biologists confront a planet populated by millions of species and their discrimination is not an easy task. The great taxonomist Carl Linnaeus introduced the binomial species nomenclature about two centuries ago, which is still in use today. This is focused mainly on morphology and a milestone towards a classification system of the species. This routine species identification has four significant limitations. First both phenotypic plasticity and genetic variability in the characters employed for species recognition can lead to incorrect identification. Secondly, this approach overlooks morphologically cryptic taxa, which are common in many groups (Knawlton 1993, Jarman and Elliott, 2000). Third, since morphological keys are often effective only for a particular life stage or gender, many individuals cannot be identified. Finally, although modern interactive versions represent a major advance, the use of keys often demands such a high level of expertise that misdiagnoses are common. The limitations inherent in morphology-based identification system and dwindling pool of taxonomists signal the need for a new approach for taxon recognition to identify the biodiversity. Not only the inherent characters of many species which flower once in their life time like bamboos, and some other organisms which flower for a day or hours make it difficult for identification but also guide the scientists to opt for a new and advanced approach. The microgenomic identification system or DNA taxonomy, which permits life's discrimination through the analysis of small segment of genome, represents one extremely promising approach to the diagnosis of biological diversity. This approach has gained popularity among the least morphologically tractable groups such as viruses, bacteria and protisis (Nanney 1982, Pace 1997, Allender *et al.,* 2001, Hamels *et al.,* 2001).

DNA sequences genomes from a uniform locality can be a barcode of life for identifying species. Biologists have used distinguishing features in taxonomic keys to apply binomial nomenclature system, for example, *Cocos nucifera.* As a master key opens all the rooms in a building, the binomial species name accesses all knowledge about a species. Now, it is well acknowledged that a short DNA sequences from a uniform locality on genomes can also be a distinguishing feature. As a Linnaean binomial is an abbreviated label for the morphology of a species, the short DNA sequence is an abbreviated label for the genome of the species. Thus the barcode of life is an additional master key to knowledge about the life forms or species.

DNA barcoding is a novel system designed to provide rapid, accurate and automatable species identifications by using short, standardized gene regions as internal sp. tags. A series of studies have now been made to test the effectiveness of DNA barcoding in species assemblages from varied geographic settings and from numerous taxonomic groups with divergent life history and evolutionary attributes. As a consequence of these sensitivity tests, barcode records are now available for more than 13,000 animal species (and accumulating rapidly) and they reveal resolution that is no illusion (www.barcodinglife.org).

According to Hebert, the pioneer in barcoding, that DNA barcode library for animals will grew by at least 500,000 records over the last 5 years, providing coverage for some 50,000 species. The barcode coverage for fishes, birds, and pest insects approaches completion which will provide open access to the identification of these species regardless of life stage or condition. Although, there has never been an effort to implement a microgenomic identification system on large scale, enough work has been done to indicate key design elements. It is clear that mitochondrial genome of animals is a better target for analysis than nuclear genome of animals, because of its lack of introns, its limited response to recombination and its haploid mode of inheritance (Saccone *et al.,* 1989). A portion of mitochondrial CO1 gene was deliberately chosen for use in animal identification when DNA barcoding was proposed (Hebert *et al.,* 2003a), and its broad utility in animal systems has been demonstrated in subsequent pilot studies (Hebert *et al.,* 2003a,b, 2004a,b and Hog and Hebert 2004). Though the taxonomic limits CO1 barcoding in animals are not fully known it has proved useful to discriminate species tested in most groups (Hebert *et al.,* 2003a).

The CO1 genes used for barcoding analysis of animal systems are not suitable for plants. Selection of gene regions and tests of their effectiveness are under progress as indicated by the results on plants (Kress *et al.,* 2005). Hence, the application of barcoding of plants is well appreciated and accorded top priority in the current and future scheme for those who handle plant germplasm at large like national museums, Botanic Gardens, seed banks, *in vitro* bank and herbaria.

DNA barcoding follows the same principle as does the basic taxonomic practice of associating a name with a specific reference collection in conjunction with a functional understanding of species concept (*i.e.*, interpreting discontitunities in interspecific variation). Though same controversies exist over the value of DNA barcoding, Besansky *et al.*, 2003, Janzen 2004, Janzen *et al.*, 2005, Hebert *et al.*, 2003 a, 2003 b, 2004 a, b and Kress 2004 have offered arguments for the utility of DNA barcoding as a powerful frame work for identifying species.

For plant molecular systematic investigation at a species level, the internal transcribed spacer (ITS) region of nuclear ribosomal cistron is the most commonly sequenced locus (Alvarez and Wendel (2003). This region has shown broad utility across photosynthetic eukaryotes (with the exception of ferns and fungi) and has been suggested as a possible plant barcode locus (Stoeckle 2003). For phylogenetic investigation, the plastid genome has been more readily exploited than the nuclear genome, and may offer for plant barcoding what the mitochondrial genome does for animals. It is uniparently inherited, non-recombining and is found structurally stable.

Recently, CBOL Plant Working Group, which included 52 researchers from 25 institutions, announced agreement on a DNA barcode for land plants (Proc Natl Acad Sci USA-2009). They have compared the performance of 7 leading candidate plastid DNA regions (*atpF–atpH* spacer, *matK* gene, *rbcL* gene, *rpoB* gene, *rpoC1* gene, *psbK–psbI* spacer, and *trnH–psbA* spacer). The broad community agreement presented here, to sequence *rbcL* and *matK* as a standard 2-locus barcode, is thus an important step in establishing a centralized plant barcode database as a tool for taxonomy, conservation, and the multitude of other applications that require identification of plant material.

The recent trend appears that DNA barcoding needs to be used along side with traditional taxonomic tools thereby problem cases can be identified and errors detected. Non-cryptic species can generally be resolved by either traditional or molecular taxonomy without ambiguity. And finally most of the global biodiversity remains unknown, molecular barcoding can only hint at the existence of new taxa, but not delimit or describe them.

References

Allender, T., Emerson, S.U., Engle, R.E., Purcell, R.H. and Bukh, J., 2001. A virus discovery method incorporating DNase treatment and its application to the identification of two bovine parvovirus species. *Proc. Natl. Acad. Sci., USA,* 98: 11609–11614.

Alvarez, I. and Wendel, J.F., 2003. Ribosomal ITS sequences and plant phylogenetic inference. *Mol. Phylogenet. Evol.,* 29: 417–434.

Besansky, N.J., Severson, D.W. and Ferdig, M.T., 2003. DNA barcoding of parasites and invertebrate disease vectors: What you don't know can hurt you. *Trends Parasitol.,* 19: 545–546. (doi:10.1016/j.pt.2003.09.015).

Hamels, J., Gala, L., Dufour, S., Vannuffel, P., Zammatteo, N. and Remacle, J., 2001 Consensus PCR and microarray for diagnosis of the genus *Staphylococcus*, species, and methicillin resistance. *BioTechniques,* 31: 1364–1372.

Hebert, P.D.N., Ratnasingham, S. and deWaard, J.R., 2003b. Barcoding animal life: Cytochrome c oxidase 1 divergences among closely related species. *Proc. R. Soc. Lond. B Biol. Sci.,* 270(Suppl).: S596–S599.

Hebert, P.D.N., Cywinska, A., Ball, S.L., and deWaard, J.R., 2003a. Biological identifications through DNA barcodes. *Proc. R. Soc. Lond. B Biol. Sci.,* 270: 313–322.

Hebert, P.D.N., Penton, E.H., Burns, J.M., Janzen, D.H., and Hallwachs, W., 2004a. Ten species in one: DNA barcoding reveals cryptic species in the neotropical skipper butterfly *Astraptes fulgerator. Proc. Natl. Acad. Sci., U.S.A.,* 101: 14812–14817.

Hebert, P.D.N., Stoeckle, M.Y., Zemlak, T.S., and Francis, C.M., 2004b. Identification of birds through DNA barcodes. *PLoS Biology,* 2: 1657–1663.

Hogg, I.D., and Hebert, P.D.N., 2004. Biological identifications of springtails (Hexapoda: Collembola) from the Canadian Arctic using mitochondrial DNA barcodes. *Can. J. Zool.,* 82(5): 749–754.

Janzen, D.H., 2004. Now is the time. *Philos. Trans R. Soc. Lond. B. Biol. Sci.,* 359: 731–732.

Janzen, D.H., Hajibabaei, M., Burns, J.M., Hallwachs, W., Remigio, E. and Hebert, P.D.N., 2005. Wedding biodiversity inventory of a large and complex Lepidoptera fauna with DNA barcoding. *Phil. Trans. R. Soc. B,* 360. (doi:10.1098/rstb.2005.1715).

Jarman, S.N. and Elliott, N.G., 2000 DNA evidence for morDNA-based identifications P.D.N. Hebert and others 02PB0653.9phological and cryptic Cenozoic speciations in the Anaspididae, 'living fossils' from the Triassic. *J. Evol. Biol.,* 13: 624–633.

Knawlton, N., 1993. Sibling species in the sea. *A. Rev. Ecol. Syst.,* 24: 189–216.

Kress, W.J., 2004. Paper floras: How long will they last? A review of flowering plants of the neotropics. *American Journal of Botany,* 91: 2124–2127.

Kress, J.W., Wurdack, K.J., Zimmer, E.A.C., Weigt, L.A. and Janzen, D.H., 2005 Use of DNA barcodes to identify flowering plants. *Proc. Natl Acad. Sci. USA,* 102: 8369–8374. (doi:10.1073/pnas.0503123102).

Nanney, D.L., 1982. Genes and phenes in *Tetrahymena. Bioscience,* 32: 783–788.

Pace, N.R., 1997. A molecular view of microbial diversity and the biosphere. *Science,* 276: 734–740.

Saccone, C., DeCarla, G., Gissi, C., Pesole, G. and Reynes, A., 1999 Evolutionary genomics in the Metazoa: the mitochondrial DNA as a model system. *Gene,* 238: 195–210.

Stoeckle, M., 2003. Taxonomy, DNA, and the bar code of life. *BioScience,* 53: 796–797.

2013, Glimpses of Animal Biodiversity   *Pages* **85–90**
Editor: **Dr. K. Muthuchelian,** *Vice Chancellor, Periyar University, Salem*
Published by: **Daya Publishing House, NEW DELHI**

Chapter 9

Biodiversity of Butterflies in Ayya Nadar Janaki Ammal College Campus, Sivakasi, Virudhunagar District

Ga. Bakavathiappan*, M. Pavaraj, S. Baskaran and L. Isaiarasu

Post-graduate and Research Department of Zoology,
Ayya Nadar Janaki Ammal College (Autonomous),
Sivakasi – 626 124

ABSTRACT

In the present study an attempt was carried out to compare the butterfly species composition and dominant families of butterflies available in Ayya Nadar Janaki Ammal College campus during 2010 with that of the earlier reported work in 1985 and 2001. During the year 1985 relative abundance of butterfly families like Papilionidae and Pieridae was high and Acraeidae, Satyridae species was low. In 2001 the relative abundance of Nymphalidae family was high and Acraeidae, Satyridae was low. In 2010 the relative abundance of Pieridae, Danaidae and Nymphalidae was high and Acraeidae was low. Three species of butterflies *viz. Colotis danae danae* (Pieridae), *Vindula erota* (Nymphalidae) and *Mycalesis perseus typhlus* (Satyridae) are newly found in the campus.

Keywords: *Butterflies, Diversity, Conservation, Relative abundance.*

* Corresponding Author: E-mail: ga.bakavathiappan@gmail.com

Introduction

Butterflies are the most attractive insects belonging to the order Lepidoptera. is one of the largest insect orders with more than 105,000 known species (Nayer *et al.,* 1990). Biodiversity of wild insects and their genetic diversity contributes to the development of agriculture, medicine and scientific research. Butterflies considered as beneficial insects and are no exception to the adverse effects of human civilization. Butterflies are the most attractive than most other insects. They are valuable pollinators, food chain components, ecological indicators and candidate material for the study of genetics, insect plant interactions and co-evolution (Sath *et al.,* 2002). Butterflies have been admired for centuries for their physical beauty and behavioral display. These colorful insects frequent open, sunny wildflower gardens, grassy fields and orchards, feeding on nectar from flowering plants. The presence of butterfly species at a particular habitat depends on a wide range of factors of which the availability of food and climatic conditions are the most important, the abundance of predators; parasitoids and the prevalence of disease also determine the abundance and density of butterfly populations. Studies on butterflies in any area would help us to understand the status of the ecosystem (Pollard and Yates, 1993). Butterflies contribute a great deal of interest and fascination to the natural world. They increase the earth's beauty. The most important environmental use is that butterflies serve as ecological indicators. They are capable of supplying information on changes in the ambient features of any ecosystem (Gunathilagaraj *et al.,* 1997). Urban biodiversity has received very little attention from conservation biologist as compared to natural and protected ecosystems (Jules, 1997). Sivakasi is one of the rain shadowed areas of Tamil Nadu that receives meager annual rainfall. However, this place witnesses a high average rainfall in the month of November which supports a majority of a shrub and herbs with comparatively few trees. The area of the study is about 64 hectares comprising in it a variety of ecological niches with the possible exception of mountains and seas. In the present study, an attempt was carried out to compare the butterfly species composition and dominant families of butterflies available in Ayya Nadar Janaki Ammal College campus Sivakasi during 2010 with that of the reported work of Isaiarasu and Alfred Mohandass (1985) and Thangadurai and Baskaran (2001).

Methods of Study

The study was carried out in Ayya Nadar Janaki Ammal College campus which is situated at an altitude of MSL + 106m, stretched to an area of 157 acres, 6 km west of Sivakasi (9°27' N, 77°49' E) in the Sivakasi–Srivilliputhur road, Virudhunagar District, Tamil Nadu. By "all out search method" the butterflies were observed from January 2010 to July 2010. The butterflies were observed and identified in the field using standard references such as, common butterflies in India (Gay *et al.,* 1992) and some South Indian butterflies (Gunathilagaraj *et al.,* 1998). However, unfamiliar species are caught by hand net, identified in the field without killing the individuals and then released into the field. In the present study, butterflies in Ayya Nadar Janaki Ammal College campus were compared with the studies of butterflies in earlier reported work of Isaiarasu and Alfred Mohandass (1985) and Thangadurai and Baskaran (2001).

Table 9.1: Comparison of earlier works and the present work on butterflies of Ayya Nadar Janaki Ammal College Campus, Sivakasi.

Sl.No.	Isaiarasu and Alfred Mohandoss (1985)	Thangadurai and Baskaran (2001)	Present Work (2010)
1.	**Family: Papilionidae**		
	Papilio polytes romulus F.cyrus	–	*Papilio polytes romulus F.cyrus*
	Papilio polytes romulus F.stichius	–	–
	Papilio demoleus demoleus	–	–
	Polydorus hector	–	–
	Graphium agamemnon	*Graphium sarpedon teredon*	–
		Pachliopta aristolochiae aristolochiae	*Pachliopta aristolochiae aristolochiae*
		Pachliopta hector	–
		Pathysa nomius nomius	–
2.	**Family: Pieridae**		
	Catopsilia pyranthe	*Catopsilia pyranthe*	–
	Catopsilia florella	–	–
	Eurema hecabe	*Eurema hecabe simulate*	*Eurema hecabe simulate*
	Anaphaeis aurota	–	–
	Ixias Spp.	*Ixias Marianne*	*Ixias Marianne*
		Catopsilia crocale	*Catopsilia crocale*
		Cepora nerissa nerissa	*Cepora nerissa nerissa*
		Catopsilia Pomona	*Catopsilia Pomona*
			Colotis danae danae
3.	**Family: Danaidae**		
	Euploea core core	*Euploea core core*	*Euploea core core*
	Danaus plexippus	–	*Danaus plexippus*
	Danaus chrysippus chrysippus	*Danaus chrysippus chrysippus*	*Danaus chrysippus chrysippus*
	Danaus limniace	–	*Danaus limniace*
		Tirumala septentrionis dravidarum	*Tirumala septentrionis dravidarum*

Contd...

Table 9.1–Contd...

Sl.No.	Isaiarasu and Alfred Mohandoss (1985)	Thangadurai and Baskaran (2001)	Present Work (2010)
4.	**Family: Acraeidae**		
	Acraea violae	Acraea violae	Acraea violae
5.	**Family: Nymphalidae**		
	Hypolimnas misippus	Hypolimnas misippus	Hypolimnas misippus
	Junonia lemonias	–	–
	Junonia almana	–	
	Junonia hierta	–	
		Hypolimns bolina jacintha	–
		Byblia ilithyia	
		Precis orithya	
		Precis iphita iphita	Precis iphita iphita
		Precis lemonias	Precis lemonias
		Precis hierta hierta	Precis hierta hierta
			Vindula erota
6.	**Family: Lycaenidae**		
	Catachrysopis cnejus	–	Everes lacturnus syntila
	Lampides boetieus	–	
		Everes lacturnus syntila	–
		Zizeeria maha ossa	
		Syntarucus plinius	Syntarucus plinius
7.	**Family: Satyridae**		
	Melanitis leda ismene	Melanitis leda ismene	Melanitis leda ismene
			Mycalesis perseus typhlus

Results and Discussion

Butterflies of Ayya Nadar Janaki Ammal College campus were studied and compared with the earlier report (1985 and 2001). In the present study 23 species of butterflies belonging to 7 families were observed; among the 7 families, 3 families were dominant such as, Pieridae (6), Danaidae (5) and Nymphalidae (5) and other 4 families are less dominant. In present study 3 butterfly species were newly found *Colotis danae danae* (Pieridae), *Vindula erota* (Nymphalidae) and *Mycalesis perseus typhlus* (Satyridae). These butterflies were not reported by earlier workers.

Table 9.2: Number of species and relative abundance of butterflies during 1985, 2001 and 2010.

Sl.No.	Family	No. of Species in ANJA College Campus			Relative Abundance (per cent)		
		1985	2001	2010	1985	2001	2010
1.	Papilionidae	5	4	2	23	16	9
2.	Pieridae	5	6	6	23	24	26
3.	Danaidae	4	3	5	18	12	22
4.	Nymphalidae	4	7	5	18	28	22
5.	Acraeidae	1	1	1	5	4	4
6.	Lycaenidae	2	3	2	9	12	17
7.	Satyridae	1	1	2	5	4	7
	TOTAL	22	25	23			

In 1985, 22 species and 7 families of butterflies were recorded in the same campus (Isaiarasu and Alfred Mohandass, 1985), and 4 families such as, Pieridae (5), Danaidae (4), Nymphalidae (4) and Papilionidae (5) were dominant. Thangadurai and Baskaran, (2001) observed 25 species belonging to 7 families of butterflies and the Nymphalidae family was found to be dominant followed by Pieridae and Papilionidae. The dominance of the family Nymphalidae is due to the fact that it contains nearly one third of known butterflies in the world. It is distributed in all over the world. These are brightly coloured butterflies with a powerful flight. Almost all male Nymphalids and some female are found of basking in the sun, many of these butterflies are attracted to dung, rotting fruit, toddy or sap and rarely flowers (Gunathilagaraj *et al.,* 1998). The abundance of Pierids can be related to the fact that, all Pierids have great affinity to flowers. In the present study Pierids visited plants of six families like Solanaceae, Compositae, Acantheceae, Caesalpiniaceae, Verbenaceae and Malvaceae. One of the most prominent features of this group is their strong migratory habit (Gay *et al.,* 1992).

Danaidae is predominantly of tropical and sub tropical distribution, with about 300 species. They are mostly insects of plains and are not found at high altitudes. They are the most abundant and conspicuous butterflies in warm low-lying areas. Most are large, brightly coloured butterflies usually brownish with black and white markings and white spotted body segments (Gunathilagaraj *et al.,* 1998). Hence, it is evident that the butterfly fauna of Ayya Nadar Janaki Ammal College campus has undergone only slight change from 1985 to 2010; the butterfly species are more or less similar. This due to the maintenance of the College campus with a good green cover full of diversified herbs and shrubs.

Acknowledgements

The authors express the profound thanks to the Management and Head of the Department of Zoology. Ayya Nadar Janaki Ammal College (Autonomous) Sivakasi for providing facilities to carry out this work.

References

Gay, T., Kehimkar, I.D. and Punitha, J.C., 1992. *Common Butterflies of India.* WWF, Oxford University Press, Bombay, p. 67.

Gunathilagaraj, K., Perumal, T.N.A., Jayaram, K. and Ganeshkumar, M., 1998. *Some South Indian Butterflies.* Nilgiri Wildlife Environment Association, Udhagamandalam, p. 274.

Isaiarasu, L. and Mohandoss, A., 1985. A report of the most common butterflies in ANJAC Campus II. *ANJAC Journal,* 5: 84–89.

Jules, E.S., 1997. Danger in dividing conservation biology and agroecology. *Conserve Biol.,* 11: 1272–1273.

Nayer, K.K., Ananthakrishnan, T.N. and David, B.V., 1990. *General and Applied Entomology.* Tata McGraw Hill Publishing Company Limited, New Delhi, p. 589.

Pollard, P. and Yates, T.J., 1993. *Monitoring Butterflies for Ecology and Conservation.* Chapman and Hall, London, p. 274.

Sathe, T.V., Nayak, D.S., Mulani, A.C., Yadav, V.S. and Bhoje, P.M., 2002. Biodiversity of butterflies of Kolhapur city, In: *Biodiversity and Environment,* (Ed.) Aravind Kumar. APH Publishing Co., New Delhi, pp. 129–135.

Thangadurai, M. and Baskaran, S., 2001. Butterflies of ANJA College Campus, Sivakasi. *M.Sc. Project Report,* Ayya Nadar Janaki Ammal College (Autonomous), Sivakasi.

2013, Glimpses of Animal Biodiversity

Pages **91–112**

Editor: **Dr. K. Muthuchelian,** *Vice Chancellor, Periyar University, Salem*

Published by: Daya Publishing House, NEW DELHI

Chapter 10

Diversity of Aquatic and Wetland Plants in Wetlands: A Case Study from West Bengal

Subir K. Ghosh

CEM Member, IUCN
Member, Wetland Subcommittee, State Biodiversity Board, West Bengal
Secretary, Society for Ecoaquaculture and
Better Environmental Development
158 NSC Bose Road, P.O. Harinavi, Kolkata – 700 148

ABSTRACT

Wetlands of West Bengal covering only about 8.5 per cent of the wetland areas (considering water bodies >100 ha) of India, provide shelter for more than 60 per cent diversity of aquatic and wetland flora. The phenology of aquatic and wetland plants is controlled by physicochemical parameters of the waterbodies. Diversity of aquatic and wetland plants is controlled by the height of the water table as well as water quality. Diversity of wetland plants of West Bengal is richest in India and is represented by more than 380 species belonging to 170 genera and 81 families (Ghosh, 1996). More than 44 species are important as food and vegetables. Eight species of emergent hydrophytes are widely exploited for rural economy. In addition, about 21 species are medicinally important and 12 species are significant for their biological filtering capacity. A recent survey implies that more than 45 species became rare, 5 species are already endangered and 6 species are under threat. In general, wetlands of the alluvial plain of the lower Ganga deltas are richest in macrophytic plant diversity in aquatic habitat due to variations in physicochemical parameters of water and bottom sediments. About 3 million people in West Bengal otherwise depend on wetland macrophytes for their subsistence livelihood support. The East Calcutta Wetlands, the only Ramsar site of West Bengal is unique for providing ecological and economical subsidy to the Metro dwellers by earning revenue through traditional waste recycling from sewage-fed fisheries as well as production of winter paddy and growing of vegetables on solid waste.

Wetland habitat of West Bengal represents 27 genera and 44 species. The wetlands of West Bengal represent 75 per cent of the fern population of the Indian subcontinent. Sea grasses belonging to six genera and two families are extensively distributed over a stretch of about 6000km coastline (Rammurthy *et al.*, 1992). Till date no sea-grass representative has been reported from West Bengal. Salt water angiosperm diversity of West Bengal coast is represented by 39 genera and 60 species of true mangrove and mangrove associates distributed in 29 families in that part of the Sundarbans that falls within the Indian territory. Among the strictly aquatic dicot of the wetland habitat of West Bengal, there are 8 families, 8 genera and 12 species. Among the strictly aquatic monocot, the wetlands of West Bengal support 36 species belonging to 19 genera and 10 families.

Due to habitat alteration and pollution as well submerged weed *Aldrovanda vesiculosa* has become extinct from its earlier habitat in the Indian subcontinent. Population of *Mimulus orbicularis*, *Isoetes coromandelina*, *Najas marina*, *Spiranthes australis*, *Utricularia striatula* have become rare in the wetlands of West Bengal. Destruction of natural habitat of *Typha elephantina*, *T. domingensis* and *Phragmites vallatoria* for industrialization and also creation of SEZ results in loss of wetland faunal diversity. Increase in salinity (27-30 ppt) of the fresh water interface (which was previously 0-1ppt). of the Sundarbans due to Aila results in mass scale destruction of fresh water habitat in coastal West Bengal and thus decline of fresh water plant and animal diversity in the region. Immediate steps should be taken to conserve these wetland macrophyte populations to conserve biodiversity of the Indian subcontinent.

Introduction

Wetlands are hotspots of biological diversity and invaluable for sustainable living. The changes within the Ramsar Convention, the shift from 'save the waterfowl' approach to a more holistic nature-human interface has significantly reflected this approach. Yet, it must be underscored that till date, international dialogue on biodiversity hotspots is restricted to terrestrial systems. It must be noted in this context that although wetlands cover only six per cent of the earth's surface (Mitsch and Gosselink, 1986), they provide habitats for about 20 per cent of the earth's total biological diversity (Gopal, 1997). This is what we know vast reservoir of the biodiversity of wetlands remains unidentified. They are among the best hosts of plants, animals and microbes. Distribution of wetland plants in West Bengal is extremely diverse due to wide variations in the hydrologic regime and physico-chemical parameters of the soil and water they live in. The present author has been working since 1987 in the wetlands of West Bengal with special reference to diversity of aquatic and wetland macrophytes from ecological perspectives. A detailed study was done in West Bengal during 1994 to 2005 regarding socio-economic importance of the wetlands plants towards subsistence livelihood support (Ghosh, 2005). This chapter represents an overview of this work in a nutshell.

Wetland Scenario in West Bengal

The natural ecosystems of West Bengal consist of freshwater marshes, flood plains, burrow peats, oxbow lakes, hill streams and mangrove swamps. In addition to these, there are different types of manmade water bodies that can be broadly divided into perennial and temporary types. The temporary wetlands can be termed as cyclical wetlands in respect of hydrology. These water bodies are locally called ponds, beels, baors, char, dighi, bheri, sarobar, bandh, haor, sayar, nayanjali etc. Part of the irrigation canals and numerous potholes (including khana, khanda) sometimes cannot be ignored from their diversity context. However, most of the ponds and beel fisheries in Gangetic West Bengal are extensively managed for fish farming and are thus not good domain for adequate aquatic plants and animals. Unmanaged waterbodies are significant in terms of biodiversity in respect of their flora and fauna. Profile of the wetland area in West Bengal has been changed in course of time due to anthropogenic pressure and unplanned urbanisation.

The wetlands in West Bengal can be broadly divided into Gangetic alluvial plains, coastal wetlands, and those of the semi-arid regions and of sub-Himalayan West Bengal. The wetlands of Gangetic alluvial plains comprise those in North and South 24 Parganas, Calcutta, Haora, Hugli, Birbhum, Nadia, Murshidabad, Maldah, parts of North Dinajpur and entire South Dinajpur. The coastal waterbodies are mostly saline in nature (salinity 1.5 ppt to 27 ppt), and are spread over East Medinipur district and certain lower pockets of South and North 24 Parganas. Wetlands of semi-arid regions fall in Bankura, West Medinipur and Puruliya districts, while those of sub-Himalayan districts are spread over parts of Dinajpur (North and South), Jalpaiguri, Koch Bihar and Darjiling districts.

Ponds and their Diversity

Ponds are by and large the most dominated waterbodies in West Bengal. Most of these ponds are monsoon-fed waterbodies and are usually dry up during post winter to summer months. Ponds are characterized by their smaller size, shallow water table and very little difference of surface and bottom waters. Saline water ponds in the coastal belts are often characterized by the presence of acid sulphate soil. These ponds are normally less significant in terms of flora and fauna.

Fresh water ponds, particularly those in the Gangetic West Bengal are rich in their floral and faunal combination. Again there are distinct variations of floral and faunal composition among different ponds even in same locality. Most of these ponds are man-made and hold fluctuating water table throughout the year. Water qualities of these ponds vary between oligotrophic to mesotrophic conditions except for the wastewater ponds where hypereutrophication is observed. Phenological behaviours of the plants grown in these ponds are also interestingly different within a small locality (Ghosh *et al.*, 1993).

Ponds and reservoirs of laterite belts hold more or less transparent water table and bottom sediments having poor nitrogen content and less organic matter. Thus submerged plants are sparsely distributed in these ponds. Floating leaf and floating stem species are more frequent in these ponds. Floral diversity of these ponds differs than their counterparts in the Gangetic alluvial plains. Variation of flowering phenology in these ponds is very sound probably due to comparatively shallow and fluctuating water table and waters rich in micronutrients.

Ponds and perennial reservoirs in the semiarid regions are unlike the ponds of laterite belts, however, they support diverse plant association and rich vegetation of submerged species where water table remain transparent soil rich in organic matters. These waterbodies are also good halt for the water birds and numerous molluscs.

Some Ecological Significance

Wetland and aquatic plants comprise highly diverse groups with a lot of hybrids, which together forms a very rich gene pool. Among these, a significant number of plants require further intensive studies for morphological, genetic and ecological points of view. Aquatic and wetland plants, including algae, water ferns and fern allies are the major producers of aquatic food chain. Large, floating-leafed plants are the settling and dwelling places for many aquatic insects, molluscs, leaches and amphibians and are the nesting and roosting grounds for birds. Wetlands in general are the abode of fishes, reptiles and mammals. Decomposed plant matter often result in degradation of water quality. A large number of aquatic microbes survive on decomposed plant litters. Aquatic plants considered as weeds in other countries do not always reach the same status in Indian conditions.

All these wetlands sustain diverse plant communities, which have definite phenological cycles with considerable variations in flowering and fruiting times in different water-bodies for the same species Ghosh *et al.,* 1993). Depth and duration of the water regimes of these different types of wetlands are also varied, and thus, also, the aquatic plants communities. Most of these wetlands are monsoon-fed, especially the seasonal ones. Potholes besides the roads and backwaters also sustain unique types of plant associations. According to variations of physico-chemical parameters of water, biotic resources of these wetlands also vary.

Production, growth and development as well as phenology of aquatic and average temperature varies from 15°C to 31°C in different water bodies in different seasons. The pH of most of the water bodies varies from 5.5 to 8.8 and in rare cases it is extended up to 10.7 (as observed in wetlands near Durgapur Steel Plants). The average salinity ranges from 0.0 to 17.0 ppt. The dissolved oxygen was lowest in one of the sewage-fed water bodies (0.81 mg/l) and highest (11.65 mg/1) in one of the highly productive water bodies. The hardness and alkalinity varies between 10.00 to 1140.00 mg/1 and 2.83 to 387.5 mg/1 respectively. The chloride content varies widely. The lowest is 10.65 mg/1 and highest is 59,995 mg/1. Sulphate concentrations in sewage-fed water bodies are also fairly high (334.63 mg/1) in comparison to mesotrophic water bodies (10.28 mg/1). Colonisation of macrophytes is more in the latter while the former type is characterised by very poor macrophytic association (Ghosh, 1994).

Wetland and aquatic plants comprise highly diverse groups with a lot of hybrids, which together forms a very rich gene pool. Among these, a significant number of plants require further intensive studies for morphological, genetic and ecological points of view. Aquatic and wetland plants, including algae, water ferns and fern allies are the major producers of aquatic food chain. Large, floating-leafed plants are the settling and dwelling places for many aquatic insects, molluscs, leaches and amphibians and are the nesting and roosting grounds for birds. Wetlands in general are the abode of fishes, reptiles and mammals. Decomposed plant matter often result in degradation of water quality. A large number of aquatic microbes survive on decomposed plant litters. Aquatic plants considered as weeds in other countries do not always reach the same status in Indian conditions.

Diversity of Wetlands Macrophytes in West Bengal

The wetlands of West Bengal, covering only about 8.5 per cent of the wetland spread in the whole country, provides more than 60 per cent of the diversity of Indian wetland vascular flora (Ghosh, 1996). West Bengal is the richest state in India in terms of wetlands floral diversity and is represented by more than 380 species belonging to 170 genera and 83 families. At least 39 families are present in wetland habitats, which cover 105 genera and 273 species are actually over-lapping in both terrestrial and wetland conditions.

Botanical Exploration in the Wetland of West Bengal

Earlier works on floral diversity in the Indian subcontinent by J D Hooker (1872-1897) included aquatic flora in the general list. Biswas and Calder (1936) first gave a detailed account of aquatic plants of the Indian subcontinent. Later, a series of records have been published on aquatic plant resources of the country. Subramanyam (1962) described morphology and distribution of 117 taxa of aquatic angiosperms in India. Later on, Deb (1975) reported the distribution and status of 144 aquatic and wetland taxa in different states of India. Lavania *et al.* (1990) compiled a list of aquatic and wetland plants of the Indian subcontinent that included 470 taxa from aquatic and semi-aquatic habitats. Apart from this, region-wise floristic survey in different parts of the subcontinent and some relevant limnological works in the Indian subcontinent has been recorded from a series of taxonomic works which includes Prain, 1903, 1905; Biswas, 1927; Agharkar and Ghosh, 1931; Biswas, 1932, 1941; Biswas, 1950; Majumdar, 1965; Mukherjee 1974; Bennet, 1979; Maji, 1981; Maji and Sikdar, 1983; Guha Bakshi, 1984; Dey *et al.,* 1988; Naskar, 1990; Mukherjee and Chattopadhyay, 1995. A detailed account of phenology, ecology, physicochemical parameters of water and bottom sediments and economic potentialities of aquatic and wetland plants of more than 100 waterbodies distributed in different districts of West Bengal was extensively explored by the present author (Ghosh, 1993, Ghosh 1994, Ghosh, 1998, Ghosh, 2000 and Ghosh, 2002). Aquatic and wetland plants studied are broadly divided in to three major categories, *viz.*, strictly aquatic, semiaquatic and tolerant wetland species having terrestrial representatives. Salt water angiosperms of the Sundarban delta is much more explored and not emphasized in this chapter. Seagrass species was not recorded during the study.

Bryophytes and Pteridophytes of Aquatic and Wetland Habitat

Riccia fluitans and *Riccia natans* are the significant representatives of bryophytes in wetlands of West Bengal. At least 8 genera and 26 species of aquatic and wetland pteridophytes distributed in 7 families in the Indian wetlands. The wetlands of West Bengal represent 6 genera and 8 species belong to 6 families which is 75 per cent of the reported aquatic and wetland pteridophytes of India and about 8.6 per cent of the world at the genus level. However, the species level diversity of aquatic and wetland pteridophytes of West Bengal is about 31 per cent of India and little more than 2 per cent of the world known population (Table 10.1). So far no gymnosperms species has been recorded from wetlands of West Bengal.

Freshwater Angiosperms

Freshwater strictly aquatic angiosperm of the world is represented by more than 141 genera and 1023 species of which the Indian subcontinent supports 54 genera and 122 species. Wetland habitat of West Bengal represents 27 genera and 44 species.

Strictly Aquatic Monocot

The Indian subcontinent supports 12 families, 30 genera and 75 species of strictly aquatic monocot. The wetlands of West Bengal support 32 species belonging to 19 genera and 10 families (Table 10.2). The Indian wetlands represent nearly 40 per cent of the world strictly aquatic families. Pontederiaceae, Lemnaceae and Hydrocharitaceae are among the dominant families.

Strictly Aquatic Dicot

The Indian subcontinent supports 9 families, 24 genera and 47 species of strictly aquatic. In the wetland habitat of West Bengal, there are 8 families, 8 genera and 12 species (Table 10.3). This is about 33 per cent at genus level and about 25 per cent at species level diversity of the strictly aquatic plants of India. The entire Nymphaeaceae family is now under threat as the rhizomes of the members of Nymphaeaceae family are randomly harvested for alternative vegetables and medicines in rural Bengal. There is no representative of the family Podostemaceae in West Bengal.

Overlapping Families in the Wetland Habitat

In West Bengal, there are about 273 species and 104 genera distributed in 39 families (Table 10.4). Orchidaceae in the Indian subcontinent is represented by two genera, *viz.*, *Spiranthes australis* and *Zeuxine strateumatica* and both the two species have been reported from the wetlands of West Bengal. Among the overlapping families, Cyperaceae is dominant, having 68 species and 11 genera, followed by Poaceae having 45 species and 24 genera and Scrophulariaceae having 27 species and 6 genera.

Due to habitat alteration and change in physico-chemical parameters of water bodies, species like *Aldrovanda vesiculosa* have either been shifted or become extinct from the wetland habitat of West Bengal. *Alternanthera philoxeroides* is an exotic weed that invades open water interface of the wetlands extensively next to *Eichhornia crassipes* in plains. *Cana* sp. belonging to the monotypic family Canaceae is found in both the wetland and terrestrial conditions in the tropics.

Table 10.1: Wetland pteridophytes and gymnosperms of West Bengal with respect to their representatives in India and World.

Sl.No.	Family	Genera					Species				
		Worldwide (No.)	India (No.)	West Bengal (No.)	% of World	% of India	Worldwide (No.)	India (No.)	West Bengal (No.)	% of World	% of India
	Aquatic and wetland Pteridophytes										
1.	Azollaceae	1	1	1	100	100	6	1	1	16.6	100
2.	Equisetaceae	1	1	0	0	0	29	1	0	0	0
3.	Isoetaceae	1	1	1	100	100	130	11	1	0.77	9.09
4.	Marsileaceae	3	1	1	33.3	100	72	7	1	1.38	14.28
5.	Polypodiaceae	1	1	1	100	100	60	1	1	0.625	100
6.	Pteridaceae (=Parkeriaceae)	2	2	2	100	100	7	3	2	28.57	66.6
7.	Salviniaceae	1	1	1	100	100	10	2	2	20	100
	Sub total No. 7	10	8	6	60	75	314	26	8	2.54	30.77
	Gymnosperms										
	Taxodiaceae	1	0	0	0	0	3	0	0	0	0
	Total No. 8	11	8	6	54.54	75	317	26	8	2.52	30.77

Table 10.2: Stricity aquatic monocot representatives of West Bengal with respect to their counterpart in India and World.

Sl.No.	Family	Genera					Species				
		Worldwide (No.)	India (No.)	West Bengal (No.)	% of World	% of India	Worldwide (No.)	India (No.)	West Bengal (No.)	% of World	% of India
1.	Alismataceae	11	5	3	27.27	60	100	7	5	20	71.4
2.	Aponogetonaceae	1	1	1	100	100	45	10	3	6.66	30
3.	Butomaceae	1	0	0	0	0	1	0	0	0	0
4.	Cymodoceaceae	5	3	0	0	0	14	5	0	0	0
5.	Hydrocharitaceae	17	8	5	29.4	62.5	100	13	5	20	38.5
6.	Juncaginaceae	3	0	0	0	0	16	0	0	0	0
7.	Lemnaceae	4	4	3	75	75	35	15	5	14.3	33.3
8.	Liliaceae	1	0	0	0	0	1	0	0	0	0
9.	Limnocharitaceae	3	2	1	33.3	50	12	2	1	8.33	50
10.	Mayacaceae	1	0	0	0	0	10	0	0	0	0
11.	Najadaceae	1	1	1	100	100	50	10	3	6	30
12.	Pontederiaceae	8	2	2	25	100	32	3	3	9.37	100
13.	Posidoniaceae	1	0	0	0	0	3	0	0	0	0
14.	Potamogetonaceae	3	1	1	33.3	100	108	6	4	3.7	66.6
15.	Ruppiaceae	1	1	1	100	100	7	1	1	14.3	100
16.	Sparganiaceae	1	0	0	0	0	19	0	0	0	0
17.	Scheuchzeriaceae	1	0	0	0	0	1	0	0	0	0
18.	Typhaceae	1	1	1	100	100	11	2	2	18.2	100
19.	Zannichelliaceae	4	1	0	0	0	6	1	0	0	0
20.	Zosteraceae	3	0	0	0	0	18	0	0	0	0
	Total = 20	71	32	19	26.76	59.37	571	75	32	5.6	42.6

Table 10.3: Strictly aquatic dicot representatives of West Bengal with respect to their counterpart in India and World.

Sl.No.	Family	Genera					Species				
		Worldwide (No.)	India (No.)	West Bengal (No.)	% of World	% of India	Worldwide (No.)	India (No.)	West Bengal (No.)	% of World	% of India
1.	Cabombaceae	2	2	1	0	0	5	2	1	0	0
2.	Ceratophyllaceae	1	1	1	100	100	4	2	1	25	50
3.	Elatinaceae	2	2	1	50	50	36	5	2	5.55	40
4.	Hippuridaceae	1	0	0	0	0	1	0	0	0	0
5.	Hydrostachyaceae	1	0	0	0	0	22	0	0	0	0
6.	Menganthaceae	5	1	1	20	100	39	8	2	5.13	25
7.	Nelumbonaceae	1	1	1	100	100	2	1	1	50	100
8.	Nymphaeaceae	6	5	1	16.6	20	65	7	2	3.07	28.57
9.	Podostemaceae	47	10	0	0	0	268	19	0	0	0
10.	Sphenocleaceae	1	1	1	100	100	1	1	1	100	100
11.	Tetrachondraceae	1	0	0	0	0	1	0	0	0	0
12.	Trapaceae	1	1	1	100	100	3	2	2	66.6	100
13.	Trapellaceae	1	0	0	0	0	2	0	0	0	0
	Total = 13	70	24	8	11.4	33.3	449	47	12	2.67	25.53

Table 10.4: Flowering plant families having representatives in wetland habitat of West Bengal with respect to their counterparts in India.

Sl.No.	Name of the Families	Genera			Species		
		India (No.)	West Bengal (No.)	% of India	India (No.)	West Bengal (No.)	% of India
1.	Acanthaceae	3	1	33.3	9	3	33.3
2.	Amaranthaceae	2	2	100	4	4	100
3.	Amaryllidaceae	1	1	100	1	1	100
4.	Apiaceae	3	3	100	3	3	100
5.	Araceae	6	5	83.3	37	9	24.32
6.	Asclepiadaceae	1	1	100	1	1	100
7.	Asteraceae	9	7	77.8	18	9	50
8.	Balsaminaceae	1	1	100	1	1	100
9.	Boraginaceae	3	2	66.6	7	3	42.85
10.	Brassicaceae	1	1	100	1	1	100
11.	Burmanniaceae	1	1	100	5	2	40
12.	Campanulaceae	1	1	100	3	2	66.6
13.	Cannaceae	1	1	100	1	1	100
14.	Caryophyllaceae	1	1	100	1	1	100
15.	Commelinaceae	4	3	75	12	10	83.3
16.	Convolvulaceae	3	1	33.3	5	2	40
17.	Cyperaceae	26	13	50	171	72	42.1
18.	Droseraceae	2	1	50	4	3	75
19.	Elatinaceae	2	1	50	5	2	40
20.	Eriocaulaceae	1	1	100	39	8	20.5
21.	Fabaceae	4	4	100	12	9	75
22.	Gentianaceae	1	1	100	2	1	50
23.	Haloragaceae	1	1	100	5	2	40
24.	Hydrophyllaceae	1	1	100	2	1	50
25.	Lamiaceae	1	1	100	4	1	25
26.	Lentibulariaceae	1	1	100	27	15	55.5
27.	Lythraceae	3	3	100	38	11	28.9
28.	Onagraceae	1	1	100	6	4	66.6
29.	Orchidaceae	2	2	100	2	2	100
30.	Poaceae	42	24	57.14	75	45	60
31.	Polygonaceae	1	1	100	7	5	71.4
32.	Ranunculaceae	1	1	100	1	1	100
33.	Rubiaceae	2	2	100	3	3	100

Contd...

Table 10.4–*Contd...*

Sl. No.	Name of the Families	Genera			Species		
		India (No.)	West Bengal (No.)	% of India	India (No.)	West Bengal (No.)	% of India
34.	Scrophulariaceae	14	6	42.85	66	25	37.8
35.	Solanaceae	1	1	100	1	1	100
36.	Sphenocleaceae	1	1	100	1	1	100
37.	Trapaceae	1	1	100	2	2	100
38.	Verbenaceae	2	2	100	2	2	100
39.	Xyridaceae	5	1	20	7	2	28.57
	Total	**157**	**103**	**65.6**	**591**	**271**	**45.85**

Diversity of Saltwater Angiosperms

In the Indian subcontinent, saltwater angiosperms are mostly dominated in the mangrove ecosystems. The total mangrove area of India is about 6560 sq km (Anonymous, 1987) of which the mangrove area of the Indian Sundarbans covers about 4267 sq km.

True mangroves or major mangrove elements, minor mangrove elements, mangrove associates including herbaceous plants and back mangroves represent mangrove flora of the Indian Sundarbans. Mangrove species of the Indian Sundarban is represented by 79 floral species distributed into 41 families and 54 genera (Table 10.5). Among these 35 species are true mangrove types (Naskar and Guha Bakshi, 1987; Mangroves of the Sundarbans, Volume 1: India, by A. B. Chaudhuri and A. Choudhury, IUCN, 1994; Naskar, 1999). The herbaceous plants of the Sunderban mangrove and mangrove-reclaimed areas are represented by 30 species belonging to 25 genera and 13 families (Ghosh *et al.,* 1990).

The Sundarbans mangrove ecosystem is also a unique corridor for the vertebrate fauna. The forests and water bodies of the Sundarbans provide dwelling places, habitats, breeding sites and roosting ground for a wide range of vertebrate species encompassing about 250 species of fishes, 8 species of amphibians, 57 species of reptiles, 161 species of birds and 40 species of mammals many of which are endangered in other parts of the world (Choudhury and Chaudhuri, 1994).

Exotic Aquatic Weeds

Exotic weeds mostly of South American origin now occupy the major portion of the open water interface of the unmanaged waterbodies in the Indian wetlands. Among these *Eichhornia crassipes* and *Alternanthera philoxeroides* are fairly common in most of the waterbodies in India and Bangladesh. In addition to these *Sagittaria montevidensis* occupies the water edges of wastewater canals. Free floating aquatic weeds like *Salvinia molesta* and *S. cucullata* are also common in the brackish water and nutrient rich waterbodies. Submerged members of the family Hydrocharitaceae

Table 10.5: List of true mangroves in the Indian sundarbans with their family and local names.

Sl.No.	Name of the Species	Family	Local Names
1.	Acanthaceae	*Acanthus ilicifolius*	*Haraguja, Sea Holly*
2.	Acanthaceae	*Acanthus volubilis*	*Lata haraguja*
3.	Myrsinaceae	*Aegiceras corniculatum*	*Khalsi*
4.	Aegialitidaceae	*Aegialitis rotundifolia*	*Satari, Tora*
5.	Meliaceae	*Amoora cucullata*	*Amur*
6.	Avicenniaceae	*Avicennia alba*	*Kala baen*
7.	Avicenniaceae	*Avicennia marina*	*Peara baen*
8.	Avicenniaceae	*Avicennia officinalis*	*Sada baen*
9.	Tiliaceae	*Brownlowia tersa*	*Lata, Bola Sundari*
10.	Rhizophoraceae	*Bruguiera cylindrica*	*Sona champa, Thushia*
11.	Rhizophoraceae	*Bruguiera gymnorrhiza*	*Kankra, Natinga*
12.	Rhizophoraceae	*Bruguiera parviflora*	*Champa, Kankra Bokul*
13.	Rhizophoraceae	*Bruguiera sexangula*	*Banduri, Kankra*
14.	Rhizophoraceae	*Ceriops decandra*	*Goran*
15.	Rhizophoraceae	*Ceriops tagal*	*Mat Goran*
16.	Fabaceae	*Cynometra ramiflora*	*Shingara*
17.	Fabaceae	*Derris trifoliata*	*Kalilata*
18.	Fabaceae	*Derris umbellatum*	*Panilata*
19.	Euphorbiaceae	*Excoecaria agallocha*	*Genwa, Blinding tree*
20.	Euphorbiaceae	*Excoecaria bicolor*	*Genwa*
21.	Sterculiaceae	*Heritiera fomes*	*Sundari*
22.	Sterculiaceae	*Heritiera littoralis*	*Sundari*
23.	Malvaceae	*Hibiscus tortuosus*	*Paras*
24.	Rhizophoraceae	*Kandelia candel*	*Goria*
25.	Combretaceae	*Lumnitzera racemosa*	*Kripa*
26.	Arecaceae	*Nypa fruticans*	*Golpata, water coconut*
27.	Arecaceae	*Phoenix paludosa*	*Hital, sea date palm*
28.	Rhizophoraceae	*Rhizophora apiculata*	*Garjan*
29.	Rhizophoraceae	*Rhizophora mucronata*	*Garjan*
30.	Sonneratiaceae	*Sonneratia apetala*	*Keora*
31.	Sonneratiaceae	*Sonneratia caseolaris*	*Keora*
32.	Tamaricaceae	*Tamarix dioica*	*Nona Jhau*
33.	Tamaricaceae	*Tamarix gallica*	*Nona Jhau*
34.	Meliaceae	*Xylocarpus granatum*	*Dhundul, Pohar*
35.	Meliaceae	*Xylocarpus mekongensis*	*Pitamari*

are mostly old-world species and regarded as serious weed in aquatic ecosystems. *Cabomba caroliniana* var. *caroliniana* is an exotic strictly aquatic submerged species introduced in India through aquarium plants. Now the plant has been naturalised in India. However, its wild population is absent in the state. Likely *Nymphaea alba*, *Victoria amazonica* and *Victoria cruziana* have been introduced in India as cultivated plants. A tentative list of threatened plants in fresh water and oligohaline habitat is depicted in Table 10.6.

Table 10.6: Some significant threatened plants in wetlands of West Bengal those need immediate steps towards conservation.

Sl. No.	Name of the Plant	Current Status
1.	*Aldrovanda vesiculosa*	This plant was last collected during 1957 from Tripura see Deb, 1975. Now probably *Aldrovanda* is extinct from West Bengal.
2.	*Caldesia oligococca*	Distribution became restricted due to alteration of physico-chemical parameter of the habitat and also due to anthropogenic pressures.
3.	*Caldesia parnassifolia*	Same as *C. oligococca*.
4.	*Drosera burmannii*	Grazing and anthropogenic pressures.
5.	*Drosera indica*	Same as *D. burmannii*.
6.	*Euryale ferox*	No natural population has been recorded from Bengal except in Maldah, where it is cultivated for commercial purpose.
7.	*Isoetes coromandelina*	Habitat modification and overconsumption by pigs in its place of origin.
8.	*Utricularia striatula*	Interspecific competitions, grazing and habitat modifications.
9.	*Najas marina*	Change in physico-chemical parameters of the habitat and anthropogenic.
10.	*Mimulus orbicularis*	Habitat modifications and change in salinity.
11.	*Spiranthes australis*	Very poorly explored taxa, and its rarity might be due to reproductive failure or unfavourable habitat.

Wetland Birds and Other Animals

More than 113 species of wetland birds are fairly common to the wetlands of West Bengal. In India, out of 1421 sites covered for midwinter waterfowl census conducted by the Wetlands International, 34 sites have been declared as internationally important sites on the basis of Ramsar criteria. Out of 41 sites in West Bengal where Asian Waterfowl Census has been done only two to three sites have fallen in this category. Presently, there is hardly any wetland in West Bengal where 20,000 waterfowls visit in regular course except Farakka Barrage in Maldah. At least 16 reptiles and 6 species of mammals are known to live in the aquatic and wetland environment of West Bengal (Source: NBSAP report, 2002).

Biodiversity of the East Calcutta Wetlands

The East Calcutta Wetlands (earlier the salt water lakes at the eastern fringes of Kolkata) is one of the 17 case study sites designated by the Ramsar Bureau for understanding the wise use of wetlands and the only Ramsar Site of West Bengal up to 19.8.2002. The East Calcutta Wetlands (22°25' to 22°40' N latitude and 88°20' E to 88°35' E longitude) is popular for its waste recycling properties. The halophytic vegetation of the Sundarbans during early 1930s was largely dominated the East Calcutta cluster of wetlands. Later on a gradual change has taken place that resulted in change of water quality from polyhaline condition to almost fresh water (see Ghosh and Sen, 1987) with a change in the profile of flora and fauna of the region (Ghosh, 2002).

During 1945, the total area of the wetlands of the eastern part of Kolkata (new name of Calcutta) was about 8097 ha out of which about 4684 ha has been converted for fish farming with city sewage. Lack of regulatory control on these wetlands and expansion of Kolkata city towards its eastern fringes have resulted in gradual encroachment of wetlands. The present area of the conservation boundary of the East Calcutta Wetlands is about 3905 ha. These clusters of sewage-fed waterbodies presently consist of 286 bheries. The existing system not only minimises the nutrient load of wastewater discharged from the Kolkata Metropolitan city, but also provides three folds of benefits (*viz.*, fish, vegetables and paddy) for the population of the city like Kolkata and its suburbs (Ghosh, 1993 and Ghosh and Furedy, 1984).

The resource recovery system of the East Calcutta Wetlands has three sections. The bheries of the East Calcutta Wetlands produces 8000 t fish yearly. The garbage farms yield daily 150 tonnes fresh vegetables and the paddy fields produces 16,000 t winter paddy in every year. These cluster of wetlands have been selected as one of the case study sites for understanding wise use concept. The resource recovery system of East Calcutta wetlands is globally accepted (see Ghosh, 1996, 1997).

Very recently about 106 aquatic and wetland species belonging to 70 genera and 36 families have been recorded from the East Calcutta Wetlands and its adjoining areas (Ghosh, 2003). Diversity of strictly aquatic representative exhibits a poor figure having 11 families, 17 genera and 22 species. Among these there is hardly any submerged vegetation in the core fishing areas, *i.e.*, within the netting zones of the bheries except the unique communities of planktonic algae. *Eichhornia crassipes, Sagittaria montevidensis* ssp. *montevidensis, Monochoria hastata, Alternanthera philoxeroides, Polygonum barbatum, Lemna aequinoctialis, Spirodela polyrhiza* are among the dominating species found in the core area in hydro phase, while *Alternanthera paronichioides, Marsilea minuta* are common in the limosal ecophase.

Till date there are some remnants of earlier saltwater flora like *Achrostichum aureum, Excoecaria agallocha* and *Acanthus ilicifolius*, however, distribution of these species are now very sparse. For periodic (at an interval of two to three years) drying up of bheries (essential as a part of sewage fed pond management), conventional removal of topsoil and liming before recharge of water, perennial vegetation is almost impossible. Some of the common hydrophytes are found to trap heavy metals from the system (see Ghosh and Ghosh, 2003).

East Calcutta Wetlands are not significant for its biodiversity value and diversity of plants and animals in the East Calcutta Wetlands is mediocre. Still it need no further mention that increase of emergent and floating leaf hydrophytes in these clusters of wetlands without disturbing the existing practice is a need for the shelter for the wild lives. Although the owners of these bheries are merely profit minded and they hardly want to compromise with the natures call, still intensive training and motivation programme is needed to motivate the owners for keeping the system more viable (see Ghosh and Ghosh, 2003).

Traditional Practices in Wetlands of West Bengal

The value of wetlands and their resources have been accepted over the ages and there has been significant international effort to conserve wetlands. The most significant aspect of wetlands is that here, humans and nature have always shared existence and exhibited interdependence.

The benefits arising out of wetlands are of broadly two categories: ecological and economical. In terms of ecological significance, wetland plants, both living or their debris, are significant to retaining the requisite carbon and methane balance of our environs and thus, maintaining green house equilibria (Denny, 1985). Wetland plants having floating or emergent leaves are considered to be an important tool in reducing global rise in temperature. Wetlands as the homes of migratory or resident avian fauna has been referred to every now and then and is well known.

The economic values accruing out of harvesting and value-adding to wetlands resources are many. Apart from conventional resources like paddy, jute, fish, prawn, crabs and molluscs, a significant number of wetland plants can be considered as non-conventional resources. Although many crops, especially vegetables, grown in the wetlands are regarded as 'subsistence agriculture', still, they are important for the rural people in managing their livelihoods.

Then, there are major and minor plant resources harvested from the wetlands. All these have significant socio-economic value. From traditional practice, mat cultivation has been commercialised in Medinipur and North 24 Parganas. Commercial mat is obtained from the culms of *Cyperus pangorei* and *Cyperus corymbosus. C. exaltatus, C. iria, Schoenoplectus grossus* etc., yield second-grade mat sticks too. About one lakh of rural people in West Bengal are dependent on the mat industry at different levels.

Commercial cultivation and management of cattails, locally known as Hogla-pati, is obtained from *Typha elephantina* and *T. domingensis*. Similarly, commercial Shola is used for making ornamental products an art intimately linked with the culture of this state and has been in existence for the last 300 years–and obtained from the stem-pith of *Aeschynomene aspera*. More than 1,00,000 people only in Southern Bengal are dependent on this business.

Makhana (*Euryale ferox*), is an aquatic cash crop of North Bihar. Edible Makhana seeds are widely used for their known nutritive values. In West Bengal, more than 900 ha area is commercially cultivated for makhana and more than 5000 people depend on this crop for their subsistence livelihood support. The seeds and powder of the seeds have known medicinal values, and are now reportedly commanding a rich market in the West. Although Makhana was common in the jheels of West Bengal during the early parts of this century, the plant has now become extremely rare in this state. Traditional cultivation of Paniphal (*Trapa natans* var. *bispinosa*) is practiced in the wetlands of 24 Parganas, Hugli, Haora, Medinipur, Bankura, Birbhum and Maldah.

There are other minor resources of the wetlands too. They include edible plants like Kalmi (*Ipomoea aquatica*), Kachu (*Colocasia esculenta*), Hingche (*Enhydra fluctuans*), Kulekhara (*Hygrophila schulli*), Shushni (*Marsilea minuta*), Thankuni (*Centella asiatica*), Shaluk (*Nymphaea nuochali, N. pubescens*), and several others yield considerable commercial benefits to the marginalised rural folk. There are also more than 20 species of medicinal plants, like Thankuni and Bramhi (*Bacopa monnieri*), shimraiya (*Nasturtium officinale*) in the wetlands of this state. In addition, more than six species of wetland plants are presently harvested for their decorative.

Management of wetland resources can go hand in hand only with the conservation of the habitats. In fact, conservation of wetland plants is directly and indirectly related to the conservation of diversity of animal lives in these habitats, as also of microbial diversity. Not much attention has been paid to the conservation and management policies for these non-conventional resources. What is also needed is to create a few wetlands in different geographical locations to conserve the typical flora, and at least of the rare ones, of that particular region.

Problems Relating to Biodiversity

Unsatisfactory socio-economic status in the Indian subcontinent leads to overuse of natural resources of wetlands. Intensive search for alternative food resource from the wetlands for sustenance has forced an alarming level of modification of physico-chemical parameters in the natural wetland habitat in the Indian subcontinent. Recent aila has results in conversion of freshwater water (almost 0 ppt salinity) wetlands to hyper saline condition (27 ppt salinity). For this a sharp decline of floristic composition has been noticed in the affected region.

This disturbance of natural wetland habitat, eutrophication, frequent change in the settlement pattern, mono-culture practice (like fisheries) for maximum profit earning and economic instability along with basic ignorance are the major driving forces for a mediocre diversity of wetland and aquatic macrophytes in the Indian subcontinent.

Unplanned urbanisation and lack of proper management in parts of the state and extension of unmanaged fishing practice in wetlands have resulted in shifting (like makhana or fox nut *i.e.*, *Euryale ferox*), even declining of aquatic plants (as in case of malakka jhanji or *Aldrovanda vesiculosa*) in West Bengal (Table 10.6). Encroachments in wetlands, eutrophication, changes in salinity in marine ecosystem, lack of national and state level wetland management policies and rapid filling of ponds have collectively resulted in disappearance of few aquatic plants and animals in the wetlands of West Bengal. The population of monotypic aquatic pteridophyte family like Isoetaceae, is now restricted in several patches in West Bengal.

Future Management

For scientific management of wetlands and waterbodies in West Bengal several steps should be taken immediately. These steps include firstly, mapping of water bodies of West Bengal by using GIS technique and compilation of a district level scientific inventory of freshwater aquatic and wetland plants and animals of West Bengal is essential. For conservation purpose district-level seed banks and herbaria of aquatic and wetland plants including scanned images for identification may be of immense helpful. At least one large perennial waterbody should be conserved in each district for providing halt for unique aquatic and semiaquatic plants of the district. Preparation of a directory of traditional practices in wetlands and proper documentation of these practices will help to create a steady database for the state.

Acknowledgements

The author likes to acknowledge Prof. K. Muthuchelian, Director Centre for Biodiversity and Forest Studies, School of Energy, Environment and Natural Resources, Madurai Kamraj University, Tamil Nadu for invitation and providing necessary arrangement for presentation of this paper in the National Conference on Biodiversity for Sustainable Development: Threats, Indicators, Climate change, Poverty allevation and targets for safeguarding Biodiversity. The author also likes to convey gratitude to Dr. Dhrubajyoti Ghosh, Chair person, IUCN, South Asia for encouraging all the times during research activities in the wetlands of West Bengal.

References

Agharkar, S.P. and Ghosh, A.K., 1931. The composition of Bengal Flora. Proc. Indian Sci. Congr., 278 pp.

Bennet, S.S.R., 1979. *Flora of Howrah District.* International Book Distributiors, Dehradun, India.

Biswas, K.P. and Calder, C.C. 1936. *Handbook of Common Water and Marsh Plants of India and Burma*, 2nd edn, New Delhi (1954, Repr. 1984).

Biswas, K.P. and Calder, C.C. 1955. *Handbook of Common Water and Marsh Plants of India and Burma*, 2nd edn, 1936. *Health Bulletin No. 24. Malaria Bureau* No. 11 (Government Place, Calcutta), 216 pp.

Biswas, K.P., 1927. Aquatic vegetation of Bengal in relation to supply of oxygen to the water. *Jr. Dept. Sci. Univ. Calcutta*, 8: 49.

Biswas, K.P., 1932. The role of aquatic vegetation in Indian waters. In: *Sir P.C. Ray Comm. Vol.,* Calcutta.

Biswas, K.P., 1941. The role of foreign plants in the economic life of Bengal. *Sci. and Cult.,* 7: 279–284.

Biswas, K.P., 1950. Study of the flora of South Calcutta with a special reference to the flora of the University College of Science Compound, Ballygunge, Calcutta. *Bull. Bot. Soc. Bengal,* Sp. Pub. No. 1.

Bor, N.L., 1960. *The Grasses of Burma, Ceylon, India and Pakistan (excluding Bambuseae).* Oxford University Press, London, U.K.

Cook, C.D.K., 1996. *Aquatic and Wetland Plants of India: A Reference Book and Identification Manual for the Vascular Plants found in Permanent or Seasonal*

Freshwater in the Subcontinent of India South of the Himalayas. Oxford University Press, Oxford, New York, Delhi, p. 1–385.

Deb, D.B., 1975. A study on the aquatic vascular plants of India. *Bull bot. Soc. Bengal,* 29(2): 155–170.

Dey, A.K., Mandal, S., Rahaman, C.H. and Dan, P.K., 1988. A contribution to the aquatic angiosperm of Santiniketan, W.B. with reference to water environment. In: *Proceedings of the Environment Management and Planing,* (Ed.) A.K. Sen. Wiley Eastern Ltd., pp. 164–174.

Ghosh S.K., 1998b. Traditional commercial practices (TCP) in wetlands in the perspective of community based wetland management in West Bengal, India. In: *Frontiers of Zoology. Proceedings of the third Refresher Course in Zoology* 6th Feb. to 28th Feb. Department of Zoology and Academic Staff College, University of Calcutta, pp. 42–43.

Ghosh S.K., 2003b. *Jalabhumir Jiban Darshan.* In: *Lok: Banglar Dighi, Jalashaya,* (Ed.) P. Sarkar, pp. 194–208. (in Bengali vernacular).

Ghosh S.K., 2003c. Conservation and management of wetlands for sustainable development in West Bengal, India. In: *Proceedings of the UGC Sponsored National Seminar on Conservation and Management of Natural Resources to Check Environmental Degradation,* (Eds.) J.N. Bagchi and A.K.M. Anwaruzzaman. pp. 143–145.

Ghosh, D. and Furedy, C., 1984. *Resource Conserving Traditions and Waste Disposal: The Garbage Farms and Sewagefed Fisheries of Calcutta Conservation and Recycling,* 7(24).

Ghosh, D. and Sen, S., 1987. Ecological history of Calcutta's wetland conservation. *Environmental Conservation,* 14(3): 219–226.

Ghosh, S.K. and Ghosh, D., 2003. Community based rehabilitation of wetlands in West Bengal, India. In: *Contemporary Studies in Natural Resource Management in India,* (Ed.) S.B. Roy *et al.* Forest F 007. Studies Series. Inter-India Publication, New Delhi, pp. 127–148.

Ghosh, S.K. and Ghosh, D., 2003. Rehabilitating biodiversity: A community based initiative in the East Calcutta Wetlands. A Communique published through WWF–India (W.B.S.O). in collaboration with British Council Division, Kolkata, 40 pp.

Ghosh, S.K. and Mitra, A., 1997. Flora and fauna of the East Calcutta wetlands. In: *Project Report of Creative Research Group.* East Calcutta Wetlands and Waste Recycling: Primary Data. Environmental Improvement Programme, Calcutta Metropolitan Water and Sanitation Authority. pp. 37–42; *Ibid.* pp. 128–149.

Ghosh, S.K. and Santra, S.C. 1995a. Biodiversity of vascular plants in Indian wetlands: An overview. Abstract: *National Symposium on Perspective in Biodiversity.* Zoological Society, Calcutta p. 25.

Ghosh, S.K. and Santra, S.C. 1996. Domestic and municipal waste water treatment by some common tropical aquatic macrophytes. *Indian Biologist*, 28(1): 47–58.

Ghosh, S.K. and Santra, S.C., 1995b. Impact of the reed vegetation of wetlands in the perspective of economic gain of rural population: A case study in West Bengal, India. *Asian Wetland News*, 8(1): 10–11.

Ghosh, S.K. and Santra, S.C., 1997. Economic benefits of wetland vegetation for rural populations in West Bengal, India. In: *Wetlands Biodiversity and Development*, (Ed.) W. Giesen. Proceedings of Workshop of the International Conference on Wetlands and Development held in Kuala Lumpur, Malaysia, and 9-13 October 1995. Wetlands International, Kuala Lumpur, p. 119–131.

Ghosh, S.K. and Santra, S.C., 1999. Phytosuccession in coastal wetlands: A case study in mangrove reclaimed areas of Sunderbans, West Bengal. In: *Sundarbans Mangal: First William Roxburgh Memorial Seminar on Sundarbans Mangal*, (Eds.) D.N. Guha Bakshi, P. Sanyal and K.R. Naskar. Naya Prakash, Calcutta, pp. 325–339.

Ghosh, S.K. and Santra, S.C., 2003. Past and present distributional records of *Makhana* (*Euryale ferox* Salisb). and its future prospect of cultivation in West Bengal, India. In: *MAKHANA*, (Eds.) R.K. Mishra, V. Jha and P.V. Dehadrai. ICAR, DIPA, New Delhi, India, pp. 1–8.

Ghosh, S.K., 1994. Studies in the ecology of aquatic plants communities in West Bengal, India. *Ph.D. Thesis*, Kalyani University, India, 501 pp. (unpublished).

Ghosh, S.K., 1997a. On the perspective of biodiversity and conservation of wetland flora in the East Calcutta wetlands, West Bengal, India. Abstract In: *9th Biennial Botanical Conference*, Dhaka, Bangladesh, Jan–8–9, 45: 25.

Ghosh, S.K., 1997b. Insect eating plants of the wetlands. *Environ.*, 2: 47–49.

Ghosh, S.K., 1998. Non-conventional Resources from wetland ecosystems. *Training Programme on Wise Use of Wetlands*, 2–6 Feb. British Council Division, Calcutta, India.

Ghosh, S.K., 1998. Wetlands. In: *Hayto Tumio Jano: A Collection of Articles on Environmental Awareness*, (Eds.) S.K. Ghosh and D.P. Chakraborty. The Eco Club, Harinavi D.V.A.S.H. School, Kolkata, West Bengal, India, pp. 24–31, (In Bengali vernacular).

Ghosh, S.K., 1998a. *A Pictorial Directory of Common Aquatic and Wetland Plants of West Bengal*. Department of Environment, Government of West Bengal, 128 pp.

Ghosh, S.K., 1999. An investigation on the autecology of an endangered wetland plant *Mimulus orbicularis* Wall. ex. Benth. (Scrophulariaceae). *Journal of Environmental Sciences*. Green Earth Foundation, Jaipur, India, 3 (1): 17–25.

Ghosh, S.K., 1999a. Diversity of wetland flora with special reference to West Bengal. In: *Proceedings of Workshop on Lake Water Quality and Hydrology Models*, (Eds). S.K. Kar and M. Bandyopadhyay. November 22–27, Department of Civil

Engineering, Indian Institute of Technology, Kharagpur. (Sponsored by Wetlands International, South Asia) pp. 1.1–1.10.

Ghosh, S.K., 1999 b. A review of Wetland Resource Models. In: *Proceedings of Workshop on Lake Water Quality and Hydrology Models*, (Eds). S.K. Kar and M. Bandyopadhyay. November 22–27, Department of Civil Engineering, Indian Institute of Technology, Kharagpur. (Sponsored by Wetlands International, South Asia) pp. 2.1–2.9.

Ghosh, S.K., 2000. Traditional commercial practices in the tropical wetlands: A case study from Eastern India. Abstract In: *Program with Abstracts, Millenium Wetland Event, Quebec–2000*, Canada, August 6–12, pp. 353.

Ghosh, S.K., 2002a. Scope of *Makhana (Euryale ferox* Salisbury) cultivation in less remunerable wetlands: A case study from West Bengal. In: *Perspectives of Plant Biodiversity*, (Ed.) A.P. Das. Bishen Singh and Mahendra Pal Singh, Dehradun, pp. 757–768.

Ghosh, S.K., 2002b. Wetland ecosystem. In: West *Bengal State Biodiversity Strategy and Action Plan: Prepared under the National Biodiversity Strategic Action Plan (NBSAP)*. The Department of Environment, Government of West Bengal and Ramkrishna Mission, Narendrapur, West Bengal, India. pp–86–106. Executed by Ministry of Environment and Forest, Government of India, Technical implementation by Kalpavriksh and Administrative Coordination by Biotech Consortium, India Ltd., Funded by Global Environmental facility through UNDP, 182 pp and Reference and Annexure (pp. 1–100).

Ghosh, S.K., 2002c. *Reclamation and Enhancement of Biodiversity of the East Calcutta Wetlands*. Project Report prepared for British Council, Calcutta implemented through WWF–India, West Bengal State Office. 61pp

Ghosh, S.K., 2003. Rehabilitation of wetland biodiversity. In: *Life Science Update*, (Eds.) G.K. Saha and S. Ray. UGC Academic Staff College, University of Calcutta, India, pp. 24–27.

Ghosh, S.K., 2003a. *Survey of Commercial Plants in Wetlands of West Bengal.* Project Report prepared for the Department of Environment, Government of West Bengal, Implemented through WWF–India, West Bengal State Office, 200 pp.

Ghosh, S.K., 2005. *Illustrated Aquatic and Wetland Plants–In Harmony with the Mankind.* Standard Literature Company, India, (225 pp. + pp–xii+ lxiv(64) colour pages (including 240 colour plates) + 17 colour plates in front cover +1 colour plate in back cover and 412 references) 301 pp.

Ghosh, S.K., Naskar, K.R. and Santra, S.C. 1990. Herbaceous flora of the tidal estuaries of the Sundarbans and their role in soil conservation. Abstract in *National Symposium of Conservation and Management of Living Resources*. Unit of Ecology and Natural Resources, USA, Bangalore, pp. 16–17.

Ghosh, S.K., Santra, S.C. and Mukherjee, P.K., 1993. Phenological studies in Aquatic Macrophytic Plants of Lower Gangetic Delta, West Bengal, India. *Feddes Repertorium (Berlin),* 104 (1–2): 93–111.

Gopal, B., 1997. Wetland and bio-diversity : How to kill two birds with one stone? In: *Wetlands Biodiversity and Development*, (Ed.) W. Giesen. Proceedings of Workshop of the International Conference on Wetlands and Development held in Kuala Lumpur, Malaysia, and 913 October 1995. Wetlands International, Kuala Lumpur, p. 18–28.

Guha Bakshi, D. N., 1984. *Flora of Murshidabad District, West Bengal, India*. Scientific Publishers, Jodhpur, India, 440 pp.

Hooker, J.D., 1872–1897. *The Flora of British India, Vols. 1–7*. Reeve and Co., Kent London.

Jha, V.N., Barat, G.K. and Jha, U.N., 1993. Nutritional evaluation of *Euryale ferox* Salisb. (Makhana). *J. Fd. Sci. Technol.*, 28(5): 325–328.

Lavania, G.S. and Gopal, B., 1990. Aquatic vegetation of the Indian subcontinent. In: *Ecology and Management of Aquatic Vegetation in the Indian Subcontinent*, (Ed.) B. Gopal. Kluwer Academic Publishers, Netherland, pp. 29–76.

Maji, S. and Sikder, J.K., 1983: Sedges and grasses of Midnapore district, West Bengal. *Jr. Econ. Tax. Bot.,* 4: 233–253.

Maji, S., 1981. Flora of Midnapore district, West Bengal. *Ph.D. Thesis*, Calcutta Univ. (Unpubl).

Majumdar, N.C., 1965: Aquatic and semi-aquatic flora of Calcutta and adjacent localities, *Bulletin of the Botanical Society of Bengal,* 19: 10–17.

Mitsch, W.J. and Gosselink, J.G., 1986. *Wetlands*. Van Nostrand Reinhold, New York, 722 pp.

Mukherjee, A. and Chattopadhyay, R., 1995. An overview of wetlands in Hoogly District. West Bengal towards an assessment of plant resources for optimum sustainable utilization 65th Ann. Sesson of the National Academy of vegetation of the Dal lake of Kashmir. *Proc. Lim. Soc. London*, 73B: 55–56.

Mukherjee, P.K., 1974. Flora of Burdwan District. *P.R.S. Thesis*, Calcutta Univ. (Unpublished).

Naskar, K.R., 1990. *Aquatic and Semi-aquatic Plants of the Lower Ganga Delta*. Daya Publishing House, New Delhi, pp. 1–408.

Newbold, C., 1993. Wetland research and survey priorities in the British Isles and their relevance to India. In: *Biodiversity Conservation: Forests, Wetlands and Deserts*, (Eds.) B. Frame, J. Victor, and Y. Joshi. Tata Energy Research Institute, New Delhi, pp. 54–67.

Prain, D., 1903. *Bengal Plants. 2 Vols,* Calcutta, 1903a (Repr. 1963).

Prain, D., 1905. The vegetation of the districts of Hughli-Howrah and the 24-Parganas. *Rec. Bot. Surv. India,* 3: 143–329.

Ramamurthy, K., Balakrishnan, N.P., Ravikumar, K. and Ganesan, R., 1992. Sea-grasses of Coromandal Coast India. Flora of India–Series 4, Botanical Survey of India, Southern Circle, Coimbatore. pp. 1–80 and 34 colour plates.

Santra, S.C. and Ghosh, S.K. , 1997. Wetland resources: Non-conventional resource evaluation. *Wetland Management Training Programme.* 17–22 March, British Council Division.

Subramaniam, K., 1962. *Aquatic Angiosperms.* Botanical Monograph 3. Council of Scientific and Industrial Research, New Delhi, 190 pp.

West Bengal State Biodiversity Strategy and Action Plan, 2002. *West Bengal State Biodiversity Strategy and Action Plan: Prepared under the National Biodiversity Strategic Action Plan (NBSAP).* The Department of Environment, Government of West Bengal and Ramkrishna Mission–Narendrapur, West Bengal, India. pp– 86–106. Executed by Ministry of Environment and Forest, Government of India, Technical implementation by Kalpavriksh and Administrative Coordination by Biotech Consortium, India Ltd., Funded by Global Environmental facility through UNDP, 182 pp + Reference and Annexure (100 pp).

2013, Glimpses of Animal Biodiversity   *Pages* **113–117**
Editor: **Dr. K. Muthuchelian,** *Vice Chancellor, Periyar University, Salem*
Published by: Daya Publishing House, NEW DELHI

Chapter 11

Biodiversity of Butterflies in a Cultivated Land Habitat of Nagamalai Area, Madurai

P. Balaji, M. Chandrasekar, A. Santhi
and M.K. Rajan
Post-Graduate and Research Department of Zoology
Ayya Nadar Janaki Ammal College (Autonomous),
Sivakasi – 626 124

ABSTRACT

Butterflies are the most attractive insects. They are valuable pollinators, when they move plant to plant in gathering nectar and also butterflies are one of the important food chain components of the birds, reptiles, spiders and predatory insects. They are good indicators of environmental quality as they are sensitive to changes in the environment. The distribution of butterflies is dependent upon the availability of their food plants. Butterflies are the first to be affected by even the slightest disturbance to their habitat so their presence or absence serves to monitor ecological changes in their habitat thus warning us about the deteriorating environment. Hence the present work was carried out on the biodiversity of butterflies in a cultivated land habitat of Nagamalai area situated in Madurai District. The study was carried out during the months from July 2002 to April 2003. In this study period, a total number of 25 species of butterflies belonging to six families was reported from the cultivated land habitat which includes Papilionidae (16 per cent), Nymphalidae (44 per cent), Pieridae (20 per cent), Danaidae (12 per cent), Acraeidae (4 per cent) and Lycaenidae (4 per cent).The analysis of data by Shannon's index revealed that the diversity was high in the cultivated land habitat in the early-post-monsoon period and Menhinick's index indicates that the cultivated land habitat rich in butterfly species in the early-post-monsoon season (December, January, February).

Introduction

Butterflies are suitable for ecological studies, as the taxonomy, geographic distribution and status of many species are relatively well known. These insects, which are mostly phytophagous, serve as primary herbivores in the food chain and also useful as pollinators of many angiosperms as many butterflies are good bioindicators of the environment, they can be used to identity ecologically important landscapes for conservation purposes. (Sudheendrakumar, *et al.,* 2000).

The number of Indian butterflies amount to one fifth of the world total of butterfly species. The Himalayan mountain range harbours the major share of the Indian butterfly diversity (Haribal, 1992). Although, only a quarter of India's butterfly diversity is represented in the Western Ghats, it has the characteristic of high alpha diversity of butterflies in certain locations (Gaonkar, 1996). However, butterflies are exceptional, are relatively well studied and documented group of insects from this mountain range (Rathinasabapathy, 1998).

The temporal dimension of the butterfly assemblages is little known from this region but for recent report from the northern Western Ghats (Kunte, 1997). Along with a checklist of butterflies from the Siruvani forests, the present chapter provides information on the seasonal pattern in their occurrence as well. (Arun, 2003).

Study Site

In this present study the biodiversity of butterflies in cultivated land areas of Nagamalai, Madurai district was studied. These agricultural lands receive water from two sources. One is seasonally from Nilayur channel and the other source is from wells. The observed area includes the cultivation of paddy and brinjal along with the green vegetations were present along the bunds of these agricultural fields. And the study period was carried out during the months from July 2002 to April 2003.

Methods

The butterflies encounted along a fixed transect route of 3 kms was recorded regularly at an interval of six days (weekly once) for a period of ten months from July 2002 to April 2003. All the butterflies recorded at a distance of 1 to 5m from the observer were recorded during the counts. The habitats selected were visited and observed between 09 00 to 16 00 hrs by sighting to study the biodiversity of butterflies. The butterflies were collected and identified by using strandard references such as "An introduction to the study of insects" by Borror and Delong and "Some South Indian butterflies" by Gunathilagaraj *et al.,* 1998. For calculating the diversity, richness and evenness indices a transect of about 500m length was marked in each habitat.

Results

In the present study, the diversity of butterflies in cultivated land habitat of Nagamalai area near Madurai was analysed. A total number of 25 butterfly species belonging to the families of Papilionidae (16 per cent), Nymphalidea (44 per cent), Pieridae (20 per cent), Danaidae (12 per cent), Acraeidae (4 per cent) and Lycaenidae (4 per cent) were observed (Table 11.1).

Table 11.1: Relative abundance of butterfly families in the cultivated land habitat.

Sl. No	Family	No. of Species and Relative Abundance (per cent)
1.	Papilionidae	4 (16 per cent)
2.	Nymphalidae	11 (44 per cent)
3.	Pieridae	5 (20 per cent)
4.	Danaidae	3 (12 per cent)
5.	Acraeidae	1 (4 per cent)
6.	Lycaenidae	1 (4 per cent)
7.	Hesperidae	—
	Total	25 (100 per cent)

Data Analysis

The analysis data by Shannon's index revealed that the diversity of butterfly was high in the cultivated land habitat in the early-post-monsoon period. Analysis of data by richness index (Menhinick's index) indicates that the cultivated land habitat rich in butterfly species in the early post-monsoon season. (December, January and February). The total number of butterfly species in cultivated land habitat is shown in Table 11.2. The species diversity indices for cultivated land habitat in different seasons is shown in Table 11.3.

Table 11.2: Total number of butterfly species in cultivated land habitat.

Sl. No	Common Name	Zoological Name	Family
1.	Lime Butterfly	*Papilio demoleus*	Papilionidae
2.	Common mormon	*Papilio polytes*	Papilionidae
3.	Crimson rose	*Polidours hector*	Papilionidae
4.	Blue mormon	*Papilio polymnestor*	Papilionidae
5.	Common castor	*Ariadne merione*	Nymphalidae
6.	Common leopard	*Atella phalanta*	Nymphalidae
7.	Danaid eggfilly	*Hypolimnas missipus*	Nymphalidae
8.	Peacock pansy	*Junonia almana*	Nymphalidae
9.	Grey pansy	*Junonia altites*	Nymphalidae
10.	Yellow pansy	*Junonia hierta*	Nymphalidae
11.	Chocolate pansy	*Junonia iphita*	Nymphalidae
12.	Lemon pansy	*Junonia lemonias*	Nymphalidae
13.	Blue pansy	*Junonia orithya*	Nymphalidae
14.	Common sailor	*Neptis hylas*	Nymphalidae
15.	Small cabbage white	*Precis rapae*	Nymphalidae
16.	Common albatross	*Appias albina*	Pieridae
17.	Mottled emigrant	*Catopsilia pyranthe*	Pieridae

Contd...

Table 11.2–*Contd...*

Sl.No	Common Name	Zoological Name	Family
18.	White orange tip	*Ixias marianne*	Pieridae
19.	Common grass yellow	*Eurema hecabe*	Pieridae
20.	Common jezebel	*Delias eucharis*	Pieridae
21.	Plain tiger	*Danaus chrysippus*	Danaidae
22.	Common crow	*Euploea core*	Danaidae
23.	Dark blue tiger	*Tirumala septentrionis*	Danaidae
24.	Tawny castor	*Acraea violae*	Acraeidae
25.	Indian cupid	*Everus lacturnus*	Lycaenidae

Table 11.3: Species diversity indices for cultivated land habitat in different seasons.

Diversity Indices	Pre-Monsoon Season (July and August)	Monsoon Season (Sep., Oct. and Nov.)	Early Post-Monsoon (Dec., Jan. and Feb.)	Late Post-Monsoon (March and April)
RICHNESS				
No.	20	22	23	17
R1	4.7413	4.7144	5.4414	4.3373
R2	2.6968	2.3723	3.0464	2.6879
DIVERSITY				
λ	4.6464	4.9247	4.01	0.1025
H'	2.8605	2.9213	2.9807	2.4572
N1	17.4714	18.5670	19.7022	11.6729
N2	21.5217	20.3055	24.9375	9.75
EVENNESS				
E1	0.9548	0.9451	0.9506	0.8673
E2	0.8335	0.8439	0.8566	0.6866
E3	0.8669	0.8365	0.8501	0.6670
E4	1.2318	0.0936	1.2657	0.8352
E5	1.2458	0.0989	1.2799	0.8198

Discussion

The present investigation is an attempt to study the biodiversity of butterflies in cultivated land habitat of Nagamalai area of Madurai district. The study has indicated that the lepidopteran fauna of this area is rich and diversified. The richness indices such as Margalef's index and Menhinick's index revealed that the cultivated land habitat is rich in butterfly species. The agricultural land were rich in vegetation which invites more number of Lepidopterans due to the availability of water in wells, keeps the place always moist, shows a very high diversity of butterflies. Farmland is considered to support a relatively impoverished butterfly fauna with a restricted occurrence along field boundaries and in other non-cultivated areas. The field boundaries are very important to butterflies, not only as breeding sites, but also as vital links between different habitats. The "butterfly plants" can be selected for gardening to conserve the Lepidopterans in a specific habitat thereby to improve pollination and the agricultural yields economically. It helps a lot to improve agriculture, horticulture and silviculture, and it increases fruit setting and seed setting.

References

Arun, P.R., 2003. Butterflies of Siruvani Forests of Western Ghats, with notes on their seasonality. *Zoos Print,* 18(2): 1003–1006.

Borror, D.J. and Delong, D.M., 1971. Collecting and preserving insects. In: *An Introduction to the Study of Insects.* Holt, Rinehart and Winston, New York, p. 680–693.

Gunathilagaraj, K., Perumal, T.N.A., Jayarajan, K. and Ganeshkumar, M., 1998. Some South Indian Butterflies, Nilgiri Wildlife and Environment Association. *Udhagamandalam,* p. 32–36.

Gaonkar, H., 1996. Butterflies of the Western Ghats, India, including Sri Lanka: A biodiversity assessment of a threatened mountain system. Unpublished report, 51pp.

Haribal, M., 1992. *The Butterflies of Sikkim and their Natural History.* Sikkim Nature Conservation Foundation, Gangtok, 217 pp.

Kunte, K.J., 1997. Seasonal Patterns in butterfly abundance and species diversity in four tropical habitats in northern Western Ghats. *Journal of Biosciences,* 22: 593–603.

Rathinasabapathy, B., Daniel, B.A. and Mathew, G., 1998. Butterflies of Coimbatore zoological park site, Anaikatty. *Zoos Print,* 13(5).

Sudheendrakumar, V.V., Binoy, C.F., Suresh, P.V. and Mathew, G., 2000. Habitat associations of butterflies in the Parambikulam wildlife Sanctuary, Kerala, India. *J. Bombay Nat. Hist. Soc.,* 97(2): 193–201.

2013, Glimpses of Animal Biodiversity   Pages 118–123
Editor: Dr. K. Muthuchelian, Vice Chancellor, Periyar University, Salem
Published by: Daya Publishing House, NEW DELHI

Chapter 12

A Preliminary List of Spider Fauna in Perumalmalai Kodaikanal Hills, Dindigul District, Tamil Nadu, India

M. Shunmugavelu[1]* and R. Ganesan[2]**

[1]P.G and Research Department of Advanced Zoology and Biotechnology,
Vivekananda College, Thiruvedagam West – 625 217,
Madurai, Tamil Nadu
[2]Department of Life Sciences, Karpagam University,
Pollachi road, Echanari – 641021, Coimbatore, Tamil Nadu

ABSTRACT

Faunal diversity surveys and conservation efforts are often concerned with charismatic species like higher vertebrates. We need to pay attention to other faunal components also like spiders because of the significant ecological role they play. Spiders have to adapt different environmental conditions. However, they seem to prefer particular zone and select a strategic line for web building in consonance with the availability of prey. Spiders are a major component of the predatory arthropod trophic level in many ecosystems. About 38,000 species of spiders belonging to 3526 genera and 109 families are known to science. Many spider species are not yet known to science, particularly in trophics. Western Ghats is one of the eco-sensitive regions in South Asia and smaller area to be surveyed. Taking the above points consideration a preliminary study was conducted to document the spider fauna in Perumalmalai Kodaikanal hills,

E-mail: *spider_msvelu@yahoo.com; **ganepost@gmail.com

Dindigul district, Tamil Nadu, India. The faunistic survey yields 25 species of spiders belonging to 21 genera and 10 families. Araneidae was the most dominant family recording 7 species belonging to 6 genera. Guild structure analysis revealed seven feeding guilds, namely orb weavers and stalkers were dominant feeding guilds of the total collection.

Introduction

Spiders are among the most diverse groups on earth, which received the seventh ranking in global diversity after the six largest insect orders. Among various arthropods, the spiders are known for their complete dependence on predation of small insects and arachnids (Coddington and Levi, 1991). Number of entomologists has acknowledged the importance of spiders as one of the major predators in regulating the pest of different crops (Patel and Pillai, 1988). Spiders have different habitats. They may be found everywhere, on dry leaves on forest floor, tall grasses, underground caves, under bark, stones, logs, near water source, mountainous areas and inside human habitats. Some spiders dig holes in the ground and make use of shallow holes for hiding. Many spiders prefer dark and shaded location with high humidity. The knowledge on diversity and distribution of spiders in India is sparse as compared to other regions of the world. The most comprehensive description yet on Indian spiders is by Tikader (1987). Thus a pressing need exists to explore spider diversity in the country. India has a very rich diversity, smaller area to be surveyed, has a tropical climate with a biodiversity hotspot, has manpower to conduct biodiversity surveys, but the best account so far (Tikader, 1987) lists 1066 spider species.

Studies on Indian spider fauna have been carried out by different workers (*e.g.* Tikader 1980; Sanyal and Tandon, 1998; Biswas and Biswas, 1992; Patel, 2002; Gajbe, 2004) in different regions of the country and documented 1035 species belonging to 240 genera under 46 families from Indian sub continent. In this chapter, we present the result of faunistic survey conducted to document the spider diversity in Perumalmalai area Kodaikanal, Dindigul district, Tamil Nadu, India.

Materials and Methods

Survey for collection and population assessment of spiders was carried out in Perumalmalai area away 10kms from Kodaikanal hills, of the taluk division of Dindigul district in the state of Tamil Nadu, India. A survey of the study area was undertaken during December 2009 to May 2010. Sampling was performed following the concept of Ritu Chauhan *et al.* (2009) from 7AM to 10AM and 5PM to 7PM during summer and 7AM to 10AM and 4PM to 6PM during winter. Spiders were collected by adapting standard sampling procedures as described below:

1. *Active Searching.* Whenever spiders were encountered, they were carefully picked without injuring them and transferred to polythene bags into tubes containing alcohol. Small spiders were collected with the help of a brush dipped in alcohol. Sedentary spiders found on the leaf blades, tree trunks and those on the webs were caught in the jar by holding it open beneath them and by tapping the spiders into it with the lid.

2. *Pit fall method*. Pitfall traps consisted of cylindrical plastic bottles of 10cm diameter and 11 cm depth (Churchill and Arthur, 1999). Six pitfall traps were laid along each transect line at an interval of 10m each. Traps were filled with preservative.

3. *Net sweeping*. Spiders were collected by sweeping the net to and pre randomly on the vegetation. During sweeping, the net was examined at regular intervals for any trapped spiders; which were immediately transferred to polythene bags.

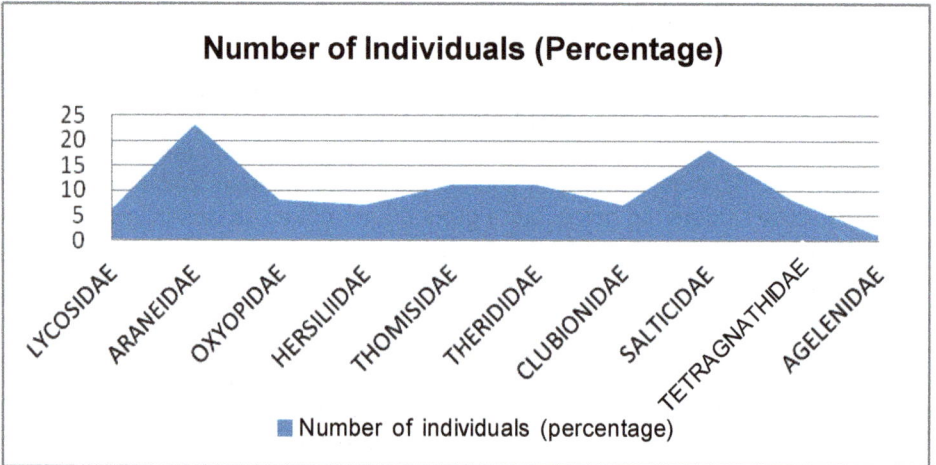

Number of Individuals (Percentage)

■ Number of individuals (percentage)

Figure 12.1: Abundance of individuals/family.

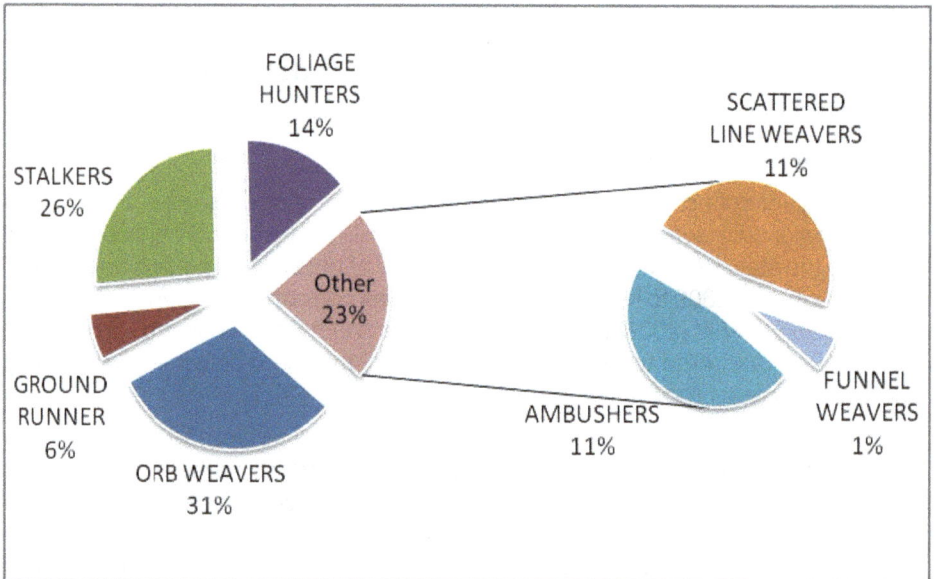

FOLIAGE HUNTERS 14%

SCATTERED LINE WEAVERS 11%

STALKERS 26%

Other 23%

GROUND RUNNER 6%

ORB WEAVERS 31%

AMBUSHERS 11%

FUNNEL WEAVERS 1%

Figure 12.2: Guild structure of spiders collected from Perumalmalai area Kodaikanal Hills.

Collected spiders were preserved in 70 per cent alcohol. Spider specimens were identified with the help of available literature (Tikader, 1987; Barrion and Litsinger 1995; Murphy and Murphy, 2000) and other literatures from the web.

Table 12.1: Spiders collected from Perumalmalai area Kodaikanal Hills.

Sl.No.	Family	Spider Fauna Genus/Species
1.	LYCOSIDAE	1. *Pardosa sumatrana* (Thorell, 1890)
2.	ARANEIDAE	2. *Araneus* sp.
		3. *Araneus nympha* (Simon, 1886)
		4. *Neoscona* sp.
		5. *Cyclosa* sp.
		6. *Gasteragantha hassetti* (CL Koch, 1837)
		7. *Argipe pulchella* (Thorell, 1881)
		8. *Neoscona mukerjei* (Tikader, 1980)
3.	OXYOPIDAE	9. *Peucetia viridana* (Stollokza, 1869)
		10. *Oxyopes birmanicus* (Thorell, 1887)
4.	HERSILIIDAE	11. *Hersillia* sp.
5.	THOMISIDAE	12. *Thomisus lobosus* (Tikader, 1965)
		13. *Thomisus* sp.
		14. *Misumena* sp.
6.	THERIDIDAE	15. *Achaearanea mundula* (L.Koch, 1872)
		16. *Chrysso* sp.
7.	CLUBIONIDAE	17. *Clubiona* sp.
8.	SALTICIDAE	18. *Myrmarachne orientales* (Tikader, 1973)
		19. *Myrmarachne* sp.
		20. *Plexippus petersi* (Karsch, 1878)
		21. *Telamonia dimidiate* (Simon, 1899)
9.	TETRAGNATHIDAE	22. *Tetragnatha* sp.
		23. *Leucauge* sp.
		24 *Leucauge decorate* (Blackwell, 1864)
10.	AGELENIDAE	25. *Agelena* sp.

Results and Discussion

Spiders representing 10 families, 21 genera and 25 species were recorded form Perumalmalai area Kodaikanal hills, Dindigul district, Tamil Nadu, India during the study (Table 12.1) period of December 2009 to May 2010. In a similar study, Tibor Magura *et al.* (2010) recorded 409 spider individuals representing 20 species in forest patches of Hungary. Araneidae was the dominant family constituting 7 species belonging to 6 genera. Abundance of different spider families in respects their individual's numbers which prominently reflects Araneidae and Salticidae as more abundant through less diverse family in comparison to Hersilidae, Clubionidae and Agelinidae (Figure 12.1).

Table 12.2: Total number spider fauna (6 months) belonging to different families and functional guild.

Sl. No.	Spider Family	Total	Guild
1.	LYCOSIDAE	40	GROUND RUNNER
2.	ARANEIDAE	153	ORB WEB WEAVERS
3.	OXYOPIDAE	55	STALKERS
4.	HERSILIIDAE	47	FOLIAGE HUNTERS
5.	THOMISIDAE	70	AMBUSHERS
6.	THERIDIDAE	70	SCATTERED LINE WEAVERS
7.	CLUBIONIDAE	46	FOLIAGE HUNTERS
8.	SALTICIDAE	117	STALKERS
9.	TETRAGNATHIDAE	52	ORB WEB WEAVERS
10.	AGELENIDAE	09	FUNNEL WEAVERS

Guild structure analysis revealed seven feeding guilds (Uetz *et al*, 1999). These are Orb weavers, stalkers, Foliage Hunters, Ambushers, Scattered line weavers, Ground runners and Funnel weavers (Table 12.2). Orb weavers constituted the dominant guild representing 31 per cent of the total collections. They are followed by Stalkers 26 per cent, Foliage Hunters 14 per cent, Ambushers 11 per cent, Scattered line weavers 11 per cent, Ground runners 6 per cent and Funnel weavers 1 per cent respectively of the total collected spider fauna (Figure 12.2). The probable reason we found behind that is the habitat type itself. Since the habitat types represent woodland, higher variety of plants and have some man-made ecosystems. It gives more opportunity to web-builders, which in turn provides the spiders with more niches and different communities like web weavers, ground runners and foliage hunters live in.

References

Barrion, A.T. and Litsinger, J.A., 1995. *Riceland Spiders of South and South East Asia.* CAB International, Wallingford England, 736 p.

Biswas, B. and Biswas, K., 1992. *Araneae: Spiders, State Fauna Series 3: Fauna of West Bengal 3.* Zool. Surv. of India, Kolkata, p. 357–500.

Chauhan, Ritu, Sihag, Vijay and Singh, N.P., 2009. Distribution and biocontrol potential of chosen spiders. *Journal of Biopesticides*, 2(2): 151–155.

Churchill, T.B. and Arthur, J., 1999. Measuring spider richness: Effects of different sampling methods and a spatial and temporal scale. *J. Insect Conserv.*, 3: 287–295.

Coddington, J. and Livi, H.W., 1999. Systematics and evolution of spiders (Araneae). *Annu. Rev. Ecol. Sys.,* 22: 565–592.

Gojbe, P., 2004. Spiders of Jabalpur, Madya Pradesh (Arachnida: Araneae). *Zoological Survey of India*, Kolkata, p. 1–154.

Magura, T., Horvath, R. and Tothmeresz, B., 2010. Effects of urbanization on ground dwelling spiders in forest patches, in Hungary. *Landscape Ecol.*, 25: 621–629.

Murphy J. and Murphy, F., 2000. An Introduction to the Spiders of south East Asia. Kola Lumpur, Malaysia. *Malaysian Nature Society*, 625 p.

Patel, B.H. and Pillai, G.K., 1988. Studies on spider fauna of Groundnut fields in Gujarat, India. *J. Bio. Control,* 2(2): 83–88.

Patel, B.H., 2002. A preliminary list of spiders with description of three new species from Parambikulam wildlife Sanctuary, Kerala. *Zoos Print, Journal,* 18(10): 1207–1212.

Sanyal, A.K. and Tandon, S.K., 1998. Arachnida. Faunal diversity in India. *Zoolo. Surv., India,* p. 349–356.

Tikader, B.K., 1980. *Fauna of India, Spiders (Thomisidae),* 1(1): 1–247.

Tikader, B.K., 1987. *Handbook of Indian Spiders.* Calcutta, Zoological Survey of India, 251 p.

Uetz, G.W., Halaj, J. and Cady, A.B., 1999. Guild structure of spiders in Major crops, *Journal of Arachnology,* 27: 270–280.

2013, Glimpses of Animal Biodiversity   *Pages* **124–132**
Editor: **Dr. K. Muthuchelian,** *Vice Chancellor, Periyar University, Salem*
Published by: **Daya Publishing House, NEW DELHI**

Chapter 13

Biodiversity of Insect Pests and their Succession in *Jatropha curcas* Conducted during 2007-08 at Raipur, Chhattisgarh

Usha Baraiha, R.N. Ganguli, Jayalaxmi Ganguli and D.J. Pophaly

Department of Entomology, College of Agriculture,
Indira Gandhi Agricultural University, Raipur, C.G.

ABSTRACT

In the studies conducted on bio-diversity of insect pests and their succession in *Jatropha curcas* at Raipur, Chhattisgarh, during 2007-08, five species of insect pests were recorded namely, blue bug, *Chrysocoris purpureus* (Westw), leaf webber cum fruit borer, *Pempelia morosalis* (Saalm uller), Coccids (unidentified), thrips, *Retithrips syriacus* and white flies along with four natural enemies namely Chrysopa, *Chrysoperla carnea*, Mantid, *Mantis* spp., two species of spiders, *Oxyopes lineatipes* and *Phidippus* spp. and Cocinellid, *Mieraspis vineta*. Natural infestation of entomophagous fungi, *Beauveria* spp. was observed in some larvae of *P. morsolis*. A Dipteran fly (unidentified) was observed in large numbers as pollinator on inflorescence of *J. curcas* plants.

The maximum population of bugs, *C. purpureus* were observed in the month of October (IInd fortnight) *i.e.* 26.98 bugs/plant and the minimum population was observed during the month of September (Ist fortnight) 9.76 bugs/plant. No population was observed during the month of April (Season II). The maximum population of larvae of webber, *P.morosalis* was observed in the month of October IInd fortnight (103.98 larvae/plant) and the minimum population was observed during the month of November IInd fortnight (34.80 larvae/plant). No population of webber was observed in the month of September in season I. The maximum population of coccids (unidentified) were observed during the month of September 1st fortnight (254.53 nymphs and adults/plant) and the minimum population in the month of October IInd fortnight (48.10 nymphs and adults/plant in season I). No population was observed during the month of November in season I and in season II. Maximum population of thrips, *Retithrips syriacus* was observed in September (141.98 larvae/plant), while minimum population (4.78 larvae/plants) in season I. No thrips were observed during season II.

Introduction

Jatropha curcas L. commonly known as "*Ratanjyot*" in Hindi belongs to the family Euphorbiaceae. It is a multipurpose large shrub or small tree, native of tropical America but thrives throughout Africa and Asia. It grows in a number of climatic zones in tropical and sub-tropical regions of the world and can be grown in areas of low rainfall and problematic sites. It is recommended as a drought resistant plant and suitable for erosion control.

Insect pests are considered to be one of the most important constraints that reduces the fruit and seed yield and oil content for which it is cultivated. In Chhattisgarh also Jatropha is attacked by number of insects among which Lepidopterous webber, scutellerid bugs and coccids are commonly found (Soman *et al.,* 2005 and Tamrakar *et al.,* 2006) which adversely effect the growth, vigour, fruiting and oil content of the plants.

In the present studies on bio-diversity five insect pests were found damaging Jatropha. Among which two were designated as major pests namely blue bug, *Chrysocoris purpureus* (Westw) and leaf webber cum fruit borer, *Pempelia morosalis* (Saalm uller), while Coccids (unidentified), thrips, *Retithrips syriacus* and white flies were recorded as a minor pests of Jatropha. Along with these, some natural enemies like Chrysopa, *Chrysoperla carnea*, Red Coccinellid beetle *Mireaspis vineta*, Mantids, *Mantis* spp., two species of spiders *viz.*, *Oxyopes lineatipes* and *Phidippus* spp., dipteran fly as pollinator and an entomophagous fungus, *Beuaveria* spp. was also observed.

Materials and Methods

The present studies on bio-diversity of insect pests and their succession was carried out at the experimental research farm, Department of Forestry, IGAU, Raipur. To study the pest succession in various provenances of *Jatropha curcas* along with natural enemies, observations were recorded fortnightly on three plants selected randomly from each provenance on the number of larvae of webber. The various species and number of natural enemies occurring were also recorded.

Results and Discussion

Blue Bug *Chrysocoris purpureus* (Westw)

The blue bug, *C. purpureus* attacked mainly the tender leaves, young shoots and the fruits of Jatropha. It was found sucking the sap from fruits, due to which the colour of fruits changed from green to yellow. In case of severe infestation the fruits turned brown. The infestation adversely affected the quality of seed and oil and ultimately reduced the fruit and oil yield. The maximum population of bugs were observed in the month of October (II fortnight) *i.e.* 26.98 bugs/plant and the minimum population were observed during the month of September (I fortnight) 9.76 bugs/plant. No population was observed during the month of April (Season II) (Table 13.1).

Webber, *Pempelia morosalis* (Saalm uller)

Webber, *P. morosalis* caused damage to the leaves, inflorescence, fruits and apical stem of the Jatropha plant by making webs along with excreta, which causes economic loss. In later stages bores into capsules, which leads reduction in oil content. The maximum population of webber larvae was observed in the month of October II fortnight (103.98 larvae/plant) and the minimum population were observed during the month of November II fortnight (34.80 larvae/plant). No population of webber larvae was observed in the month of September (Table 13.1) in season I. In season II the maximum and minimum population was in the month of April 1st fortnight (91.75 larvae/plant) and June 1st fortnight (2.56 larvae/plant), respectively.

Coccids (Unidentified)

They are the soft bodied insects. The nymphs and adults suck the plant sap from the leaves leading to yellowing, browning and curling, which has a negative impact on the growth of plants. It secretes a waxy secretion over its body. It excretes honey dew on which sooty mould develops. The maximum population of coccids were observed during the month of September 1st fortnight (254.53 nymphs and adults/plant) and the minimum population were observed in the month of October IInd fortnight (48.10 nymphs and adults/plant in season I). No population was observed during the month of November (Table 13.1) in season I (Table 13.1). and in season II.

Thrips, *Retithrips syriacus* (Mayaet)

Damage symptoms of discoloration and white speckles appear due to feeding of thrips, *Retithrips syriacus* (Mayaet) on Jatropha. Maximum population of thrips was observed in September (141.98 thrips/plant), while minimum population (4.78 thrips/ plants) in season I. No population of thrips was observed during season II. The thrips, *R. syriacus* (Mayaet) was reported to attack on *J. curcas* in Puerto Rico (Medina and Franfuis, 2001). It was recorded for the first time in India on *J. curcas* and are polyphagous in nature and found to colonize on grapevine, cotton, groundnut, castor, pomegranate cashew and rose plants (Regupathy *et al.,* 2003).

White Flies (Unidentified)

Both nymphs and adult of white flies were found sucking sap from the leaves and young shoots of Jatropha. The population was very scarce.

Table 13.1: Fortnightly mean population of insect pests along with natural enemies recorded on *Jatropha curcas*.

Months	Fortnightly Mean Population							
	Blue Bug (C.purpureus)	Webber (P.morosalis)	Coccids (Unidentified)	Thrips (Retithrips syriacus)	Spider (Oxyopes lineatipes)	Mantid (Mantis spp.)	Chrysopa (C. carnea)	Coccinellid (Mireaspis vineta)
Season-I								
Sept. 1st FN	9.769	0.000	254.538	0.000	25.979	20.907	2.997	16.984
Sept. 2nd FN	10.438	0.000	121.021	4.777	13.764	12.174	4.884	10.767
Oct. 1st FN	11.768	0.000	166.317	47.436	6.216	10.767	5.106	11.211
Oct. 2nd FN	26.983	103.982	48.101	64.321	7.437	12.396	3.441	6.327
Nov. 1st FN	23.536	78.658	0.000	141.983	7.548	14.467	1.554	12.101
Nov. 2nd FN	18.652	34.801	0.000	327.982	11.211	10.989	0.444	6.216
Season-II								
April 1st FN	0.000	91.753	0.000	0.000	4.995	13.876	0.000	0.000
April 2nd FN	0.000	69.610	0.000	0.000	9.435	14.727	0.000	0.000
May 1st FN	0.000	46.651	0.000	0.000	14.431	13.358	0.000	0.000
May 2nd FN	0.000	14.327	0.000	0.000	13.987	10.620	0.000	0.000
June 1st FN	0.000	2.553	0.000	0.000	10.434	10.545	0.000	0.000
June 2nd FN	0.000	0.000	0.000	0.000	14.099	10.878	0.000	0.000

Natural Enemies

Chrysopa, *Chrysoperla carnea*

Chrysopa, *C. carnea* was recorded as a predator feeding on soft bodies insects. The maximum population of this predator was observed during the month of October 1st fortnight (5.10 Chrysopa/plant) and the minimum population were observed during the month of November IInd fortnight (0.44 Chrysopa/plant) in season I. No population was observed during season II.

Red Coccinellid Beetle, *Mireaspis vineta*

The red Coccinellid beetle, *Mireaspis vineta* was found feeding on the coccids. Maximum population of coccinelid (16.68 beetles/plant) during I[st] fortnight of September while minimum (6.21beeltes/plant) was recorded in the 2nd fortnight of November.

Spider, *Oxyopes lineatipes* and *Phidippus* spp.

Spider, *Oxyopes lineatipes* (Figure 13.1) was seen prey mostly on moths, within striking distance. They play an important role by killing 2-3 moths daily thus preventing new generation of pest built up. The maximum population of this spider was recorded during the month of September 1st fortnight (25.97 spider/plant) in season-I and during May (14.72 spider/plant) in season-II. Minimum population was observed during the month of October 6.62 and April 4.99 spider/plant in season I and II, respectively.

Mantids, *Mantis* sp.

Two species of preying mantids, *Mantis* sp. were recorded as a predators feeding on soft bodied insects. Maximum population of these predators was observed in the 1st fortnight of September (16.43 mantids/plant) and minimum in the 1st fortnight of October (12.73 mantids/plant) in season-I, while in season II maximum population was observed in the IInd fortnight of April (14.78 mantids/plant) and minimum in IInd fortnight of June (10.54 mantids/plant).

Dipteran Fly (Unidentified)

These were observed in large numbers as pollinators on inflorescence of *J. curcas* plants.

Entomophagous Fungi, *Beauveria* spp.

Some larvae of webber, *P. morosalis* were found naturally infested by the entomophagous fungus, *Beauveria* spp. The larvae showing this type of infestation were stiff, mummified and covered with white fungus all over the body.

Conclusion

Thus, from the present studies we can conclude that the biodiversity of insect pests in Jatropha is very vast, which include few major and some minor pests along with a large fauna of natural enemies including many species of insect pests and entomophagous fungi which can be further exploited as bio-control agents. The dipteran fly as pollinator is a very important component and having a lot of scope for the future in enhancing fruit set and yield.

Figure 13.1

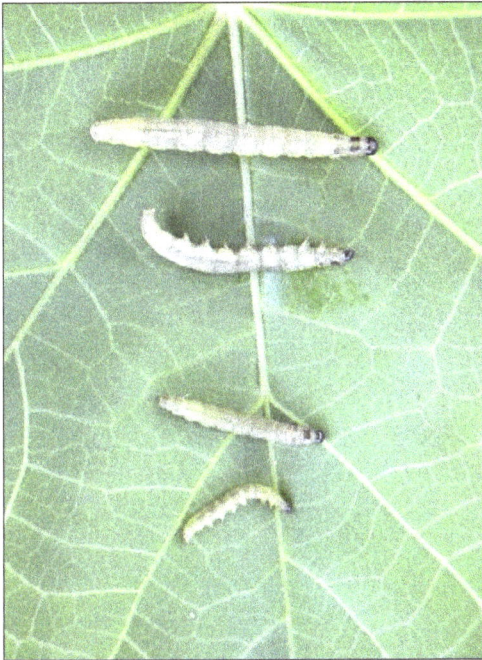

Various larval instars of webber, *P. morosalis*

Nymphs and adults of sucking sap from fruits of Jatropha, *Chrysocoris purpureus*

Various species of preying mantis recorded on *J. curcas*

Figure 13.1–*Contd...*

Thrips Retithrips syriacus larvae on leaf of *J. curcas*

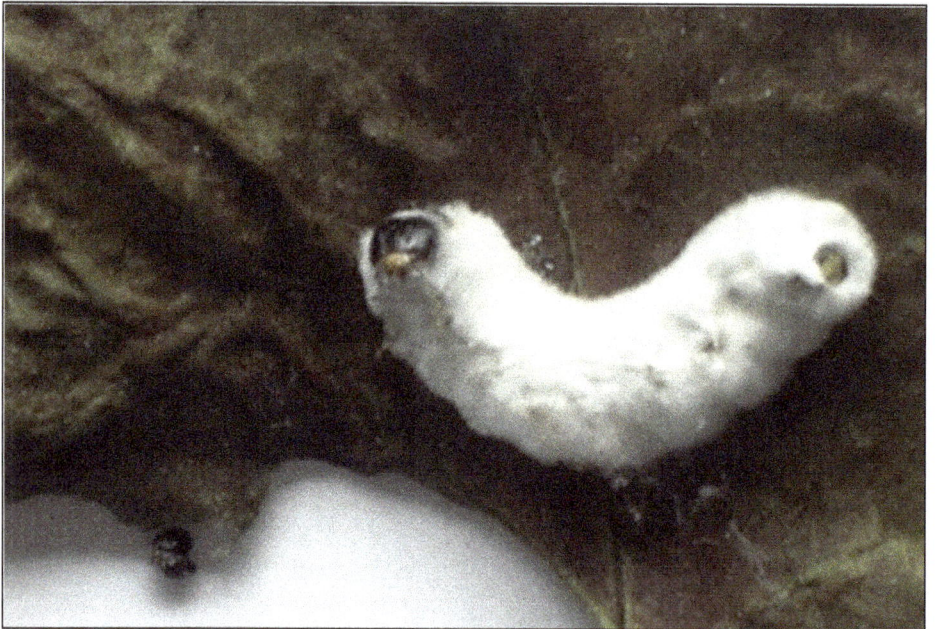

Larva of webber attacked by entomophagous fungi, *Beauveria* sp.

Figure 13.1–*Contd...*

Two species of spiders, *Phidippus* sp. and *Oxyopes lineatipes*

Dipterous fly (unidentified) acting as pollinator in flowers of *J. curcas*

References

Medina, G.S. and Frngui, R.A., 2001. *Retithrips syriacus* (Mayet) the black Vine thrips insect (Thysenoptera Thripidae) new to Puerto Rico. *Journal of Agriculture of the University of Puerto Rico*, 85(1–2): 85–89.

Reghupathy, A., Palanisamy, S., Chandramohan, N. and Gunathilagaraj, K., 2003. *A Guide to Crop Pests*. Sheeba Printers, Coimbatore, India, 276p.

Soman, D., Ganguli, J. L. and Ganguli, R. N., 2006. Bio-diversity of insect pests of multitier agroforestry system. Paper presented in *National Conference on Forest Biodiversity, Resources: Exploitation, Conservation and Management*, held at CBFS, Madurai Kamraj University, Madurai form 21–22 March, pp. 113–114.

Tamrakar, E., Ganguli, J.L., Soman, D. and Ganguli, R.N., 2007. Studies on pest succession on Jatropha (*Jatropha curcas*) under multitier agroforestry system. In: *National Symposium on Sustainable pest Management for Safer Environment*, held at OUAT, Bhubaneshwar from 6–7 Dec.

2013, Glimpses of Animal Biodiversity
Editor: Dr. K. Muthuchelian, Vice Chancellor, Periyar University, Salem
Published by: Daya Publishing House, NEW DELHI

Pages 133–146

Chapter 14

Seasonal Diversity of Mosquito Fauna in Srivilliputhur Taluk, Virudhunagar District

K. Karuppasamy* and S. Baskaran

Post-graduate and Research Department of Zoology,
Ayya Nadar Janaki Ammal College (Autonomous), Sivakasi – 626 124

ABSTRACT

Systematic survey of diversity of mosquito species of Srivilliputhur taluk have been carried out in post-monsoon, late post monsoon, premonsoon and monsoon seasons of 2000–2001. A total of 42 mosquito species belonging to 6 genera were recorded in 16 villages of Srivilliputhur taluk during the study period. Twenty-five and Twenty-six species belonging to six genera were recorded during the post monsoon and late post monsoon seasons respectively. In premonsoon 12 species belonging to 4 genera were recorded. During monsoon season 26 mosquito species belonging to five genera were recorded during the study period. Intrageneric diversity was observed in *Culex, Ades, Anophles* and *Monsonia* in post monsoon, late post monsoon and monsoon seasons. The *Monsonia* species was recorded during the post monsoon and late post monsoon seasons only.

Introduction

Public health scientists are increasingly discovering that the recent emergence or re-emergence of infectious diseases has an origin in environmental change (Eisenberg *et al.,* 2007). A common feature of emerging infectious diseases is that

_____

* Corresponding Author: E-mail: karuppasamy.anjac@gmail.com

they are associated with anthropogenic changes to the environment (Patz *et al.,* 2004; MA, 2005). Mosquito is one among the wide range of vectors, playing an important role in the transmission of several kinds of dreadful diseases to variety of animals, predominantly human beings by feeding on blood of infected organism and transferring to the uninfected human beings (Schafer, 2004). Mosquitoes are delicate insects with soft skin and slender body which belong to the order Diptera. There are currently more than 3000 mosquito species in the world which are grouped into 39 genera and 135 sub-genera. They are grouped into three sub-families, Tuxorynchitinae, Culicinae and Anophelininae (Roy and Brown, 1970). Mosquitoes can be annoying, serious problem in man's domain. Their attacks on farm animals can cause loss of weight and decreased milk production. Some mosquitoes are capable of transmitting diseases, such as, malaria, yellow fever, dengue fever, filariasis, encephalitis and even chickengunea (Reinert, 2001). The influence of the environmental conditions produced large inter and intra-specific variations among the mosquitoes. These variations in the heterogeneous environment enabled mosquitoes to develop different adaptations which ultimately cause change in mosquito behaviour (Suwonkerd *et al.,* 1995). The survey of mosquitoes provides valuable information on density, occurrence, distribution, prevalence and species composition of various mosquitoes in an area, which indirectly presumes significance due to their participation in public health importance (Prakash *et al.,* 1998). The knowledge of the presence of mosquito species at a particular time can help to predict the danger of a viral disease spread by that mosquito species. This would help to take preventive measures in advance (Sangdeep *et al.,* 1994).

Materials and Methods

Study Area

Srivilliputhur Taluk is located in Virudhunagar district in the 77°30'0" E to 77°45'0" E altitude and 9°31'0" N to 9°39'0" N latitude with elevation 823m above MSL. This is a typical rural ecosystem with an environmental heterogeneity. It is present in the foothill of Western ghats.

It receives high rainfall through northeast monsoon. Agriculture is the primary work of the people of the study area. The climate is alternately humid and dry in the year. The climatic seasons are logically divided into four main seasons such as pre–monsoon (June, July and August); monsoon (September, October and November); early post-monsoon (December, January and February) and late post-monsoon (March, April and May) seasons based on rainfall.

Study Period

The study was conducted in 16 selected study sites of Srivilliputhur Taluk. The study was carried out for fifteen months from October 2000 to December 2001. The study period covered all four seasons, such as, early post-monsoon (December 2000 to February 2001), late post-monsoon (March 2001 to May 2001), pre-monsoon (June 2001 and August 2001) and monsoon (September 2001 to November 2001) seasons. Monthly collections have been done in all the selected sites of the study area. The preserved mosquitoes were identified upto species level by the Entomologists of the

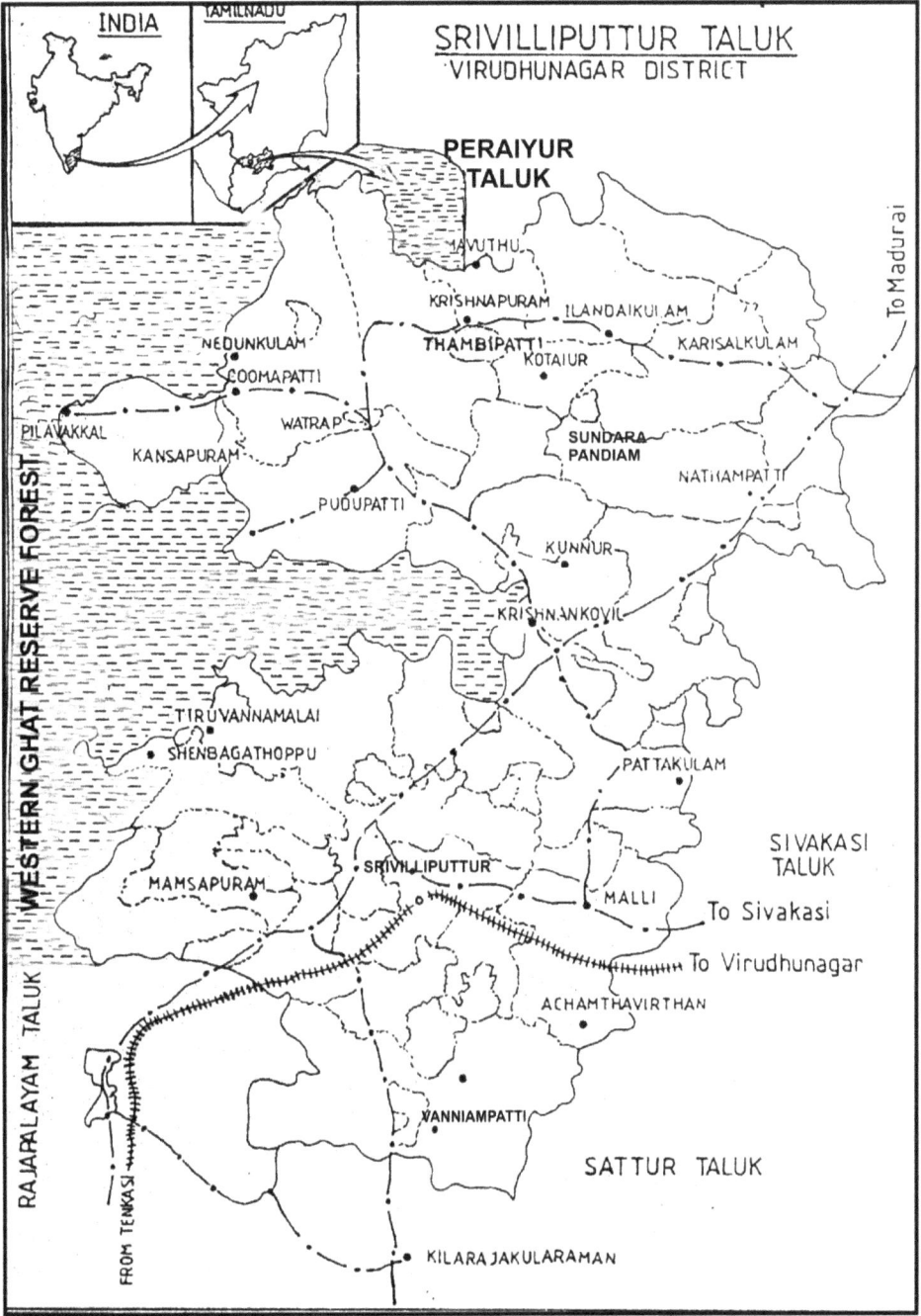

Figure 14.1: Study Sites in Srivilliputhur Taluk.

Regional Entomological Research Centre at Virudhunagar. Species of mosquito fauna of the selected study area were analysed.

Results and Discussion

A total of 42 species of mosquitoes belonging to six genera namely *Armigeres*, *Culex*, *Aedes*, *Anopheles*, *Mansonia* and *Megarhinus* were collected and recorded from 16 different study sites in Srivilliputhur Taluk, Virudhunagar District, Tamil Nadu between October 2000 and December 2001(Table 14.1). Diversity of mosquitoes recorded during the study was categorized under four seasons. They are, post-monsoon (December 2000 to February 2001), late post-monsoon (March 2001 to May 2001), pre–monsoon (June 2001 to August 2001) and monsoon (September 2001 to November 2001) seasons (Tables 14.1 to 14.4; Figures 14.2 to 14.5).

Twenty-five and twenty-six mosquito species belonging to 6 genera, were recorded during post-monsoon (11504 mosquitoes) and late post-monsoon seasons (6312

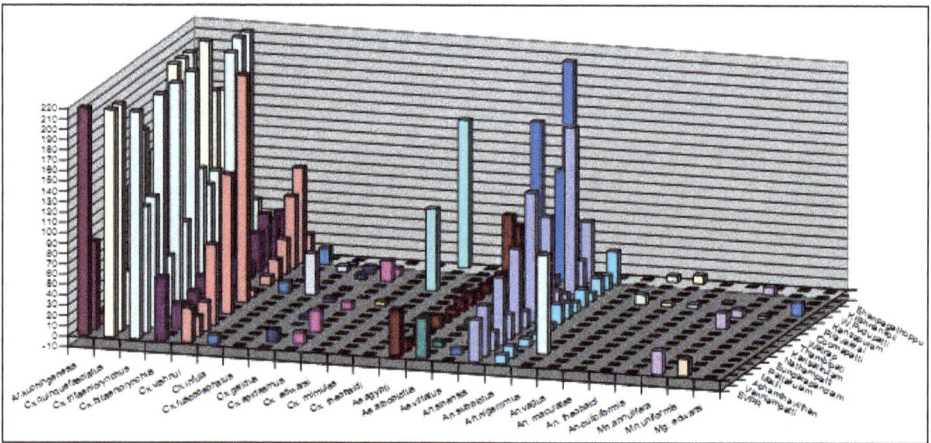

Figure 14.2: Diversity and distribution of mosquito species recorded during post-monsoon (Dec. 2000 to Feb. 2001).

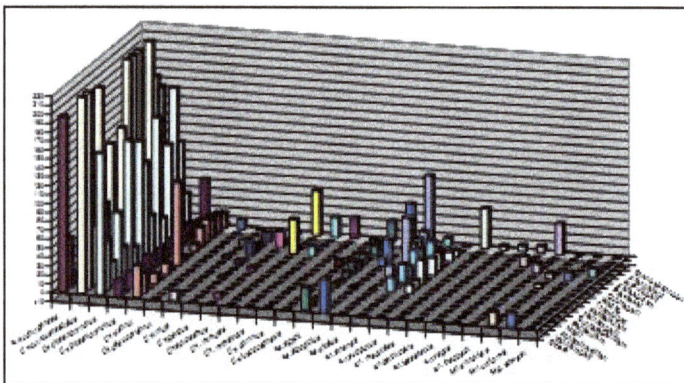

Figure 14.3: Diversity and distribution of mosquito species recorded during late post-monsoon (Mar. 2001 to May 2001).

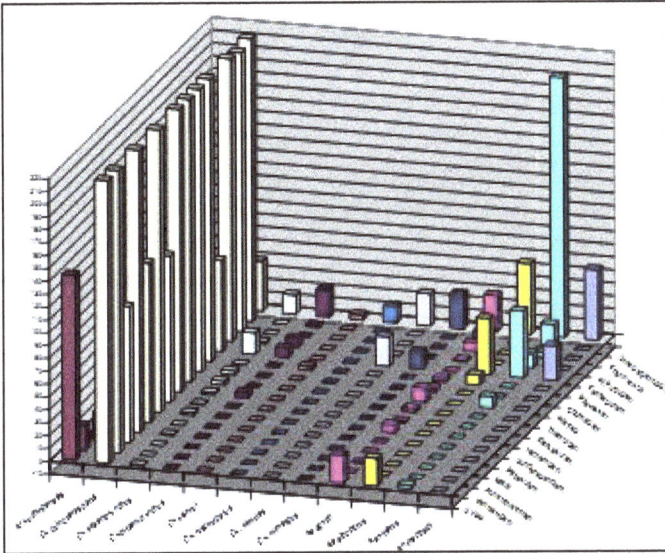

Figure 14.4: Diversity and distribution of mosquito species recorded during pre-monsoon (Jun. 2001 to Aug. 2001).

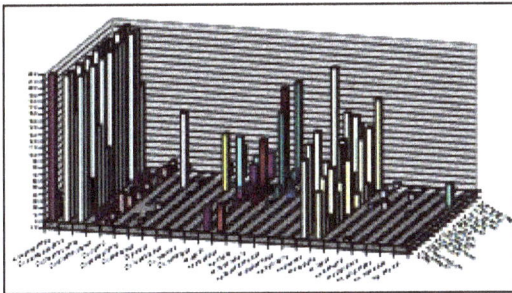

Figure 14.5: Diversity and distribution of mosquito species recorded during monsoon (Sept. 2001 to Nov. 2001).

mosquitoes) respectively (Tables 14.3 and 14.4). In pre-monsoon, 12 species with least number of 4661 mosquitoes belonging to 4 genera were recorded (Table 14.3). Twenty-six mosquito species with the maximum number of 13194 mosquitoes were recorded under 5 genera during the monsoon season (Table 14.4). Intra-generic diversity was observed in *Culex, Aedes, Anopheles* and *Mansonia* in post-monsoon, late post-monsoon and monsoon seasons. *Armigeres* and *Megarhinus* were represented by single species each. *An. theobaldi* was the only species belonging to the genus *Anopheles* recorded in pre-monsoon season. The genus *Mansonia* was recorded during the post-monsoon and late post-monsoon seasons only.

The diversity of mosquito species in the monsoon (26 species), post-monsoon (25 species) and late post-monsoon (26 species) was found to be higher when compared with pre-monsoon period where less number (12 species) of species was collected (Tables 14.1 to 14.4 and Figures 14.2 to 14.5). The reason for the maximum

Table 14.1: Distribution of different mosquito species recorded in the chosen study sites of Srivilliputhur Taluk during post-monsoon season (from Dec. 2000 to Feb. 2001).

Sl.No.	Name of the Mosquito Species	Collection Sites and No. of Mosquito Species																Total
		SRI	VAN	ACH	MAL	PAT	SUN	NAT	KAR	THA	WAT	COO	PIL	KAN	WPU	KRI	SHE	
1.	Ae. aegypti	45	11	6	13	3	12	6	8	11	12	8	9	8	9	60	45	266
2.	Ae. albopictus	39	–	6	–	–	2	4	–	–	–	5	–	–	–	3	–	59
3.	Ae. vittatus	6	6										170	11	10	106	206	515
4.	An. maculates												4				8	12
5.	An. theobaldi												4					4
6.	An. culiciformis													2				2
7.	An. nigerrimus	–	–	96														96
8.	An. sinensis	41	49	19	65	24	82	16	125	61	91	14	6	160	27	59	–	839
9.	An. subpictus	9	12	–	8	7	22	6	19	15	10	17	8	17	14	32	1	197
10.	An. vagus												10				6	16
11.	Ar. kuchingenesis	340	86	10	10	14	–	–	25	90	25	19	–	23	–	20	–	662
12.	Cx. edwarsi									4								4
13.	Cx. mimules												79				143	222
14.	Cx. theobaldi												2					2
15.	Cx. bitaeniorynchus	61	1	25	20	26	35	17	18	9	37	25	16	38	51	41	43	463
16.	Cx. epidesmus	–	11	–	21	–	–	–	7	–	–	–	–	20	7	2	–	66
17.	Cx. fuscocephalus	–	–	–	–	–	–	–	–	–	40	–	–	–	6	2	–	48
18.	Cx. gelidus		12				3	2		4	1	2		5	7	–		36
19.	Cx. infula	9	–								11			8		16	–	44
20.	Cx. quinquefasciatus	252	494	53	130	179	108	148	47	646	337	380	20	423	167	162	38	3584

Contd...

Table 14.1–*Contd...*

Sl.No.	Name of the Mosquito Species	Collection Sites and No. of Mosquito Species																Total
		SRI	VAN	ACH	MAL	PAT	SUN	NAT	KAR	THA	WAT	COO	PIL	KAN	WPU	KRI	SHE	
21.	*Cx. tritaeniorynchus*	263	123	127	331	59	512	81	451	121	100	108	18	211	635	286	53	3479
22.	*Cx. vishnui*	33	20	28	78	–	135	–	253	7	–	9	20	34	70	92	22	801
23.	*Mg. edwarsi*												12					12
24.	*Mn. annulifera*	23								16	–	8				8		55
25.	*Mn. uniformis*	17										3						20
	Total																	**11504**

Table 14.2: Distribution of different mosquito species recorded in the chosen study sites of Srivilliputhur Taluk during late post-monsoon season (from Mar. 2001 to May 2001).

Sl.No.	Name of the Mosquito Species	Collection Sites and No. of Mosquito Species																Total
		SRI	VAN	ACH	MAL	PAT	SUN	NAT	KAR	THA	WAT	COO	PIL	KAN	WPU	KRI	SHE	
1.	Ae. aegypti	27	4	4	7	–	5	3	8	14	9	12	–	11	10	6	23	143
2.	Ae. albopictus	37					3					9	24				44	117
3.	Ae. vittatus									13			51	17	17	20	77	195
4.	An. maculates																45	45
5.	An. theobaldi												11				35	46
6.	An. barbirostris																5	5
7.	An. sinensis						17	5	21	14	26	23	–	21	9	–	9	145
8.	An. subpictus						10	10	27	–	20	20	–	11	8			106
9.	An. tessellatus																7	7
10.	An. vagus													12			8	20
11.	Ar. kuchingenesis	197	56	8	9	7	–	–	14	57	14	25	–	–	3	9	–	399
12.	Cx. corntus																23	23
13.	Cx. mimeticus												10				22	32
14.	Cx. mimules												39				50	89
15.	Cx. bitaeniorynchus	19		17	9	14	20	13	16	13	17	–	19	19	15	12	53	256
16.	Cx. epidesmus		6						5					19				30
17.	Cx. fuscocephalus										9							9
18.	Cx. gelidus						3			5		14	2	4	10			38
19.	Cx. infula	11								3				5				19
20.	Cx. pseudovishnui																11	11

Contd...

Table 14.2–*Contd...*

Sl.No.	Name of the Mosquito Species	Collection Sites and No. of Mosquito Species																			Total
		SRI	VAN	ACH	MAL	PAT	SUN	NAT	KAR	THA	WAT	COO	PIL	KAN	WPU	KRI	SHE			
21.	*Cx. quinquefasciatus*	230	273	66	215	81	142	53	149	319	287	341	60	256	176	159	19			2826
22.	*Cx. tritaeniorynchus*	159	56	27	77	34	147	23	136	107	43	147	29	106	164	99	36			1390
23.	*Cx. vishnui*	36	22	–	12	–	13	–	93	5	–	8	22	25	23	22	17			298
24.	*Mg. edwarsi*												10							10
25.	*Mn. annulifera*	16								5	–	7								28
26.	*Mn. uniformis*	15										10								25
	Total																			**6312**

Table 14.3: Distribution of different mosquito species recorded in the chosen study sites of Srivilliputhur Taluk during pre-monsoon season (from June 2001 to Aug. 2001).

Sl.No.	Name of the Mosquito Species	Collection Sites and No. of Mosquito Species																Total
		SRI	VAN	ACH	MAL	PAT	SUN	NAT	KAR	THA	WAT	COO	PIL	KAN	WPU	KRI	SHE	
1.	*Ae. aegypti*	22					6	1	3	11	3	3	–	1	7	8	28	93
2.	*Ae. albopictus*	22										8	44				55	129
3.	*Ae. vittatus*	3								9			52	11	4	18	202	299
4.	*An. theobaldi*												27				55	82
5.	*Ar. kuchingenesis*	144	19	3	13	7	–	–	10	45	12	17	12	–	–	7	–	289
6.	*Cx. fuscifurgatus*																13	13
7.	*Cx. mimetigus*												13				27	40
8.	*Cx. mimules*												23				25	48
9.	*Cx. bitaeniorynchus*								4			7	7	7		1	21	40
10.	*Cx. quinquefasciatus*	309	309	106	224	124	256	114	274	325	287	321	71	351	217	247	39	3574
11.	*Cx. tritaeniorynchus*	1	–	–	1	–	2	1	2	3	3	–	17	2	–	–	15	47
12.	*Cx. vishnui*											3					4	7
	Total																	**4661**

Table 4: Distribution of different mosquito species recorded in the chosen study sites of Srivilliputhur Taluk during monsoon season (from Sep. 2001 to Nov. 2001).

Sl.No.	Name of the Mosquito Species	Collection Sites and No. of Mosquito Species																Total
		SRI	VAN	ACH	MAL	PAT	SUN	NAT	KAR	THA	WAT	COO	PIL	KAN	WPU	KRI	SHE	
1.	*Ae. hubernatoris*											14						14
2.	*Ae. pallirostris*															7		7
3.	*Ae. aegypti*	38	2	3	3	2	–	4	10	23	12	18	42	15	8	6	39	225
4.	*Ae. albopictus*	43								18	21	86	11	4	–	138		328
5.	*Ae. albotaeniatus*																6	6
6.	*Ae. pseudotaeniatus*												1			19		20
7.	*Ae. vittatus*									6		18	123	7	10	21	148	339
8.	*An. culicifascies*									4								4
9.	*An. maculates*									2								2
10.	*An. uliciformis*													10				10
11.	*An. inensis*	118	135	18	144	36	89	37	225	97	100	136	39	121	101	87	11	1494
12.	*An. stephensi*	5									13							18
13.	*An. subpictus*	71	88	26	91	48	61	–	82	–	22	98	25	104	–	137	–	853
14.	*An. vagus*										6	2	–	12	–	7	–	27
15.	*Ar. kuchingenesis*	316	37	5	12	13	16	–	18	129	20	29	–	29	2	4	–	630
16.	*Cx. fuscifurgatus*															17	17	17
17.	*Cx. mimetigus*												83			33	33	116
18.	*Cx. mimules*												86			47	47	133
19.	*Cx. bitaeniorynchus*	–	17	–	11	5	8	–	4	–	5	22	–	12	–	–	–	84
20.	*Cx. epidesmus*	–	5	–	5	–	8						113					131

Contd...

Table 14.4—Contd...

Sl.No.	Name of the Mosquito Species	Collection Sites and No. of Mosquito Species																Total
		SRI	VAN	ACH	MAL	PAT	SUN	NAT	KAR	THA	WAT	COO	PIL	KAN	WPU	KRI	SHE	
21.	Cx. gelidus										4			6	2			12
22.	Cx. pseudovishnui															10	10	10
23.	Cx. quinquefasciatus	247	355	156	286	165	461	148	383	548	408	533	124	434	356	399	43	5046
24.	Cx. tritaeniorynchus	306	338	54	258	124	292	81	299	290	317	361	–	305	311	134	–	3470
25.	Cx. vishnui	6	13	7	32	–	17	–	12	9	14	–	–	7	17	13	16	163
26.	Mg. edwarsi												35					35
	Total																	**13194**

diversity in the study area during monsoon, post and late monsoon seasons may be due to the occurrence of more temporary and permanent water bodies for the oviposition of female mosquitoes and larval inhabiting. The same result was recorded in the study undertaken in Kenya where the rainfall enhanced the diversity of mosquitoes and its larval forms (Hutchings *et al.*, 2005). The population of this species was associated with the cultivation of paddy. Kanojia *et al.* (2003) from Gujarat showed a bimodal pattern of Japanese encephalitis vector, *Cx. tritaeniorhynchus* population with short and tall peaks during dry month (March) and tall during rainy months (August to October). Seasonal variation of mosquito abundance is a rather well–documented pattern and can often be associated with weather and hydrological conditions. Schafer (2004) reported seasonal variations of mosquito species in Dalaven region in Sweden.

The decreased number of species in the pre-monsoon may be due to dry condition and reduced breeding sites in the study sites. According to Adebote *et al.* (2008) when the number of breeding sites are more in the rainy and less after rainy months, the diversity of mosquitoes varied. Low mosquito diversity was reported in the dry season of 2002 in 2 villages of Gedaref state, New Jersy (Chevalier *et al.*, 2004). This result merges with the finding of the present study. The present findings also coincide with the report of Muturi *et al.* (2006) in Mwea, Kenya where the diversity was more during the cultivation season. The species of mosquito found at any one time in an area is dependent on temperature and seasons. The diversity and distribution of mosquito species showed variation in different seasons such as, early spring, late spring, summer and mid summer in Northern eastern United States (Cranes, 2004). Diversity of mosquito species varied in different study sites and study period. Varied diversity and distribution of mosquito species in different study sites in the study period was due to the availability of many mosquitogenic habitats and variety of domestic, non-domestic and human host for breeding and feeding in different study period. Similar results were reported by Schafer (2004) in Sweden and Cranes (2004) in Massachusetts. Forty two mosquito species recorded in Srivilliputhur Taluk in different seasons indicated the diversity and richness of mosquito species. The heterogeneity of inter and intra-generic variation among the mosquito species is mainly due to the availability of suitable hosts and appropriate breeding habitats for phytophilic, zoophilic and anthropophilic mosquitoes.

References

Adebote, D.A., Oniye, S.J. and Muhammed, Y.A., 2008. Studies on mosquitoes breeding in rock pools on Inselbergs around Zaria, northern Nigeria. *J. Vector Borne Dis.*, 45: 21–28.

Chevalier, S., de la Rocque, S., Baldet, T., Vial, L. and Roger, F., 2004. Epidemiological processes involved in the emergence of vector-borne diseases: West Nile fever, Rift Valley fever, Japanese encephalitis and Crimean-Congo haemorrhgic fever. *Rev. Sci. Tech. Off. Int. Epiz.*, 23(2): 535–555.

Cranes, J.W., 2004, A classification system for mosquito life cycles: Life cycles types for mosquitoes of North eastern United States. *J. Vector. Ecol.*, 1: 1–10.

Eisenberg, J.N.S., Desai, M.A., Levy, K., Bates, S.J., Liang, S., Naumoff, K. and Scott, J.C., 2007. Environmental determinants of infectious disease: A framework for tracking causal links and guiding public health research, *Environ. Health. Perspect.*, 115: 1216–1223.

Hutchings, R.S., Sallum, M.A., Ferreira, R.L. and Hutchings, R.W., 2005, Mosquitoes of the Jau National Park and their potential importance in Brazilian Amazonia, *Med. Vet. Entomol.*, 19 (4): 428–441.

Kanojia, P.C., Shetty, P.S., and Geevarghese, G., 2003. A long-term study on vector abundance and seasonal prevalence in relation to the occurrence of *Japanese encephalitis* in Gorakhpur district, Uttar Pradesh. *Ind. J. Med. Res.*, 117: 104–110.

Kulkarni, S.M. and Naik, P.S., 1989. Breeding habitats of mosquito in Goa. *Ind. J. Malariol.*, 26(1): 41–44.

MA (Millennium Ecosystem Assessment), 2005. *Ecosystems and Human Well-Being: Current State and Trends.* Island Press, Washington, DC.

Muturi, E.J., Shililu, H., Gu, W., Jacob, B.G., Githure, J.I. and Novk, R.J., 2006. Mosquito species diversity and abundance in relation to land use in a rice land agro-ecosystem in Mwea, Kenya. *J. Vector. Ecol.*, 31(1): 129–137.

Patz, J.A., Daszak, P., Tabor, G.M., Aguirre, A.A., Pearl, M. and Epstein, J., 2004. Unhealthy landscapes: Policy recommendations on land use change and infectious disease emergence. *Environ. Health Perspect.*, 112: 1092–1098.

Prakash, A., Bhattacharya, D.R., Mohapatra, P.K. and Mahanta, J., 1998. Mosquito fauna in a broken forest ecosystem of District Dibrugarh with updated systematic list of mosquitoes recorded in Assam, India. *Entomon.*, 23(4): 291–298.

Reinert, J.F., 2001, Revised list of abbreviations for genera and subgenera of Culicidae (Diptera) and notes on generic and subgeneric changes. *J. Am. Mosq. Cont. Associ.*, 17: 51–55.

Sagandeep, V., Kapoor, C. and Grewal, J.S., 1994, Some mosquito (Diptera : Culicidae) species of Punjab and Himachal Pradesh. *J. Insect. Sci.*, 7(1) : 48–50.

Schafer, M., 2004. Mosquitoes as a part of wetland biodiversity, comprehensive summaries of Uppasala dissertations from the faculty of Science and Technology, Uppasala University, p. 63.

2013, Glimpses of Animal Biodiversity   *Pages* **147–157**
Editor: **Dr. K. Muthuchelian,** *Vice Chancellor, Periyar University, Salem*
Published by: Daya Publishing House, NEW DELHI

Chapter 15

Impact of Monsoon on the Density of Mosquito Fauna in Srivilliputhur Taluk, Virudhunagar District

K. Karuppasamy* and S. Baskaran

*Post-graduate and Research Department of Zoology,
Ayya Nadar Janaki Ammal College (Autonomous), Sivakasi – 626 124*

ABSTRACT

Seasonal density of different mosquito species was carried out in 16 villages of Srivilliputhur taluk from December 2000 to November 2001. A total of 35671 mosquitoes belonging to 42 species under six genera were recorded during the study period. Among the total mosquitoes 11504, 6312, 4661 and 13194 mosquitoes were recorded in post monsoon, late post monsoon, pre monsoon and monsoon seasons respectively. *Culex quinquefasciatus* showed highest density in all the seasons. The mosquito Culex *triaeniorhychus showed* highest density in all the seasons expect pre-monsoon. *Armigeros kuchingenesis* and *Culex vishnui* exhibited moderate density in all the four seasons *Culex bitaeniorhynchus* density was moderate in post monsoon and late post monsoon seasons. Among the *Aedes* species *Aedes aegypti, Aedes albopictus* and *Aedes vittatus* exhibited moderate density in all the four seasons. The density of *Anopheles sinensis* was high in post monsoon, late post monsoon and premonsoon season. *Anopheles subpictus* showed moderate density in all the three seasons except premonsoon. The remaining mosquito species exhibited less density in all the four seasons.

* Corresponding Author: E-mail: karuppasamy.anjac@gmail.com

Introduction

Mosquitoes are one of the largest vectors of diseases globally. Currently, over 3000 species of mosquitoes have been identified worldwide, a number of which are known animal and human disease vectors. Mosquito-borne diseases pose a major threat both to the diversity of indigenous fauna and to human populations throughout the world. The number of individuals of each species of mosquito fauna available in a particular area determine the density. When a particular place contains thick density means, the environmental conditions, breeding sources and blood meal host availability are favourable for the mosquitoes. Density of mosquito fauna determines some important vector-borne diseases. Vector-borne diseases emerge mainly due to climate change. There are two mechanisms by which climate change may impact vector borne diseases. One mechanism of change is brought about by the direct impact of climate change on the agent, vector, or host. For instance, changes in temperature, rainfall, humidity, or storm patterns may directly affect the multiplication or differentiation rate of the vector or the agent, the biting rate of the vector, or the amount of times that the host is exposed to the vector. A second mechanism of environmental change is brought about by the indirect impacts of climate. In this category, climate influences some parameters that are important to vector spread or survival, such as the type of agriculture or vegetation, which, in turn, changes the relationship between the parasite, the vector, and the host (Lopes *et al.,* 2005). Rain, irrigation for cultivation, water storage for human utilization purposes and other biotic and abiotic factors also influence the density of mosquito population of different species. For effective management of mosquitoes, ecology, lifecycle, behaviour as well as faunistics of mosquitoes still requires indepth investigation (Pandian and Manoharan, 1994). Since density of mosquitoes plays major role in determining outbreak of vector-borne diseases, adequate knowledge about density of mosquito species is necessary to eradicate the vectors. The density of each species has its favourable factors that may influence on abundance. Therefore, the present study has been designed to evaluate the density of the various mosquito species available during different months and seasons in all the 16 study sites.

Materials and Methods

Human bait collections, resting collections and larval collections were made to assess the density of mosquitoes in the study area by using the method adopted by Pandian and Chandrashekaran (1980). Density of mosquitoes was determined by collecting the adult female mosquitoes by human bait resting mosquitoes from their roosting places and larval collections. The number of each mosquito species collected in all 16 sites were recorded properly, tabulated and analysed by various possible calculations in order to find out the various factors responsible for three different degrees of density like dominant, sub-dominant and satellite pattern and correlated with seasons and other mosquitogenic conditions during the study period.

Results and Discussion

Seasonal density of different mosquito species was studied in the study area from December 2000 to November 2001. A total number of 11504, 6312, 4661 and

13194 mosquitoes was recorded in post monsoon/winter, late post monsoon/summer, pre monsoon and monsoon/rainy seasons respectively (Tables 15.1 to 15.4). The density of different mosquito species of the four seasons is shown in Tables 15.1 to 15.4 and Figures 15.1 to 15.4. *Culex quinquefasciatus* showed highest density in all the seasons. The mosquito species *Culex tritaeniorhynchus* showed highest density in all the seasons except pre monsoon. *Armigeres kuchingenesis* and *Culex vishnui* exhibited moderate density in all the four seasons and *Culex bitaeniorhynchus* density was moderate in post-monsoon and late post-monsoon seasons and showed less density in pre-monsoon and monsoon seasons. Among the *Aedes* mosquitoes *Aedes aegypti, Aedes albopictus* and *Aedes vittatus* exhibited moderate density in all the seasons (Tables 15.1 to 15.4; Figures 15.1 to 15.4). The density of *Anopheles sinensis*

Table 15.1: Density and frequency of different mosquito species recorded in Srivilliputhur Taluk during post-monsoon season (Dec. 2000–Feb. 2001).

Sl. No.	Name of the Species	Total	Density (per cent)	Frequency (per cent)
1.	Ae. aegypti	266	2.31	100
2.	Ae. albopictus	59	0.51	37.5
3.	Ae. vittatus	515	4.48	43.75
4.	An. theobaldi	4	0.035	6.25
5.	An. culiciformis	2	0.017	6.25
6.	An. maculatus	12	0.104	12.5
7.	An. nigerrimus	96	0.83	6.25
8.	An. sinensis	839	7.29	93.75
9.	An. subpictus	197	1.71	93.75
10.	An. vagus	16	0.14	12.5
11.	Ar. kuchingenesis	662	5.75	68.75
12.	Cx. infula	44	0.38	25
13.	Cx. mimulus	222	1.93	6.25
14.	Cx. theobaldi	2	0.017	6.25
15.	Cx. bitaeniorhynchus	463	4.02	100
16.	Cx. edwarsi	4	0.035	6.25
17.	Cx. epidesmus	66	0.57	31.25
18.	Cx. fuscocephalus	48	0.42	18.75
19.	Cx. gelidus	36	0.31	50
20.	Cx. quinquefasciatus	3584	31.15	100
21.	Cx. tritaeniorhynchus	3479	30.24	100
22.	Cx. vishnui	801	6.96	81.25
23.	Mg. edwarsi	12	0.104	6.25
24.	Mn. annulifera	55	0.48	25
25.	Mn. uniformis	20	0.17	12.5
	Total	**11504**		

was high in post-monsoon, late post-monsoon and pre-monsoon seasons. *Anopheles subpictus* showed moderate density in all three seasons except pre-monsoon. The remaining mosquito species exhibited less density in all the four seasons.

The density of different mosquito species in the study period varied considerably over seasons. High density of *Culex quinquefasciatus* was recorded throughout the study period. High environmental variability in combination with a wild climate is likely to be the cause for high mosquito population. This may be due to the unplanned human inhabitations and poor drainage facilities leading to sewage stagnation for the breeding of the mosquitoes. Further, availability of high human population in the urban

Table 15.2: Density and frequency of different mosquito species recorded in Srivilliputhur Taluk during late post-monsoon season (March 2001–May 2001).

Sl.No.	Name of the Species	Total	Density (per cent)	Frequency (per cent)
1.	*Ae. aegypti*	143	2.27	87.5
2.	*Ae. albopictus*	117	1.85	31.25
3.	*Ae. vittatus*	195	3.09	37.5
4.	*An. barbirostris*	5	0.08	6.25
5.	*An. theobaldi*	46	0.73	12.5
6.	*An. maculatus*	45	0.71	6.25
7.	*An. sinensis*	145	2.29	56.25
8.	*An. subpictus*	106	1.68	43.75
9.	*An. tessellatus*	7	0.11	6.25
10.	*An. vagus*	20	0.32	12.5
11.	*Ar. kuchingenesis*	399	6.32	68.75
12.	*Cx. cornotus*	23	0.36	6.25
13.	*Cx. infula*	19	0.30	18.75
14.	*Cx. mimulus*	89	1.41	12.5
15.	*Cx. bitaeniorhynchus*	256	4.06	87.5
16.	*Cx. epidesmus*	30	0.48	18.75
17.	*Cx. fuscocephalus*	9	0.14	6.25
18.	*Cx. gelidus*	38	0.60	37.5
19.	*Cx. mimeticus*	32	0.51	12.5
20.	*Cx. pseudovishnui*	11	0.17	6.25
21.	*Cx. quinquefasciatus*	2826	44.78	100
22.	*Cx. tritaeniorhynchus*	1390	22.02	100
23.	*Cx. vishnui*	298	4.72	75
24.	*Mg. edwarsi*	10	0.16	6.25
25.	*Mn. annulifera*	28	0.44	18.75
26.	*Mn. uniformis*	25	0.40	12.5
	Total	**6312**		

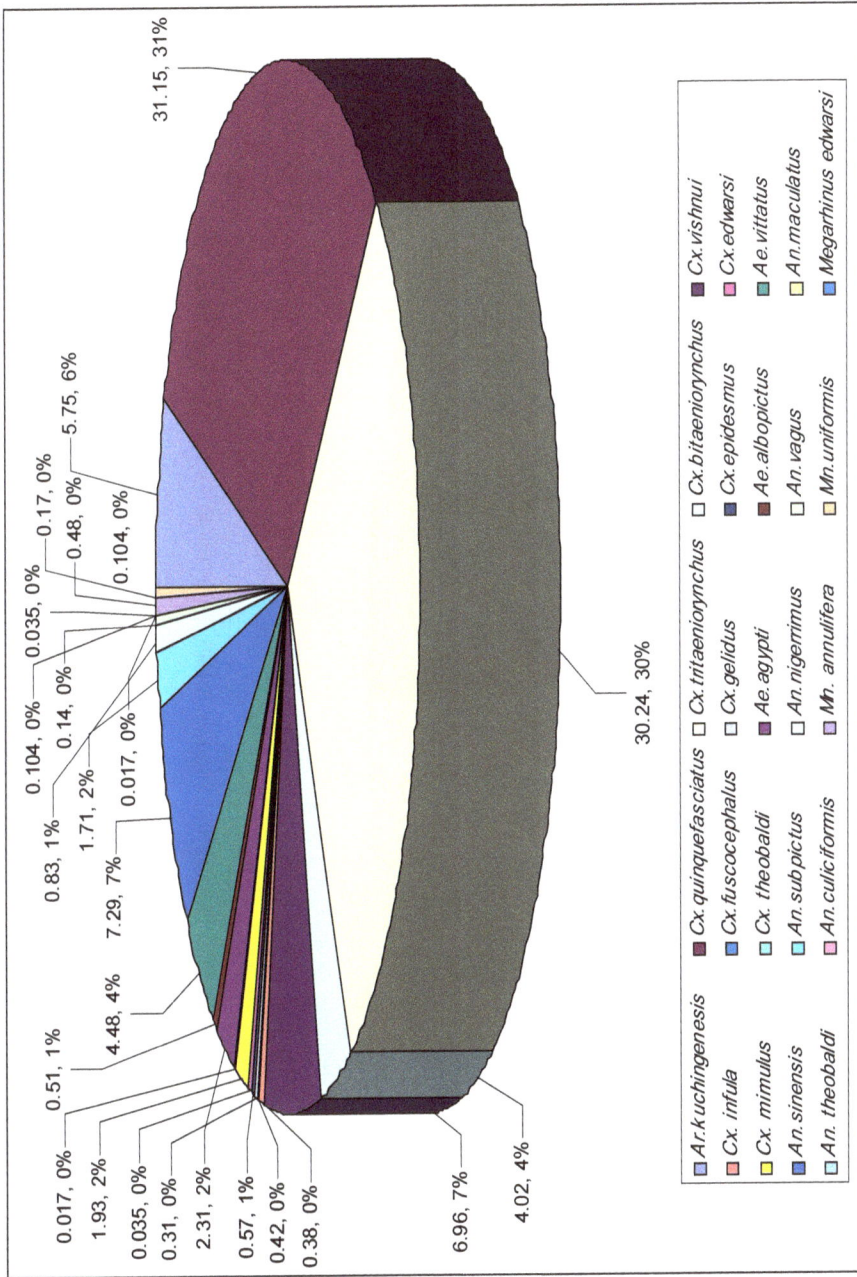

Figure 15.1: Density of different mosquito species recorded in Srivilliputhur Taluk during post-monsoon Season (December 2000 to February 2001).

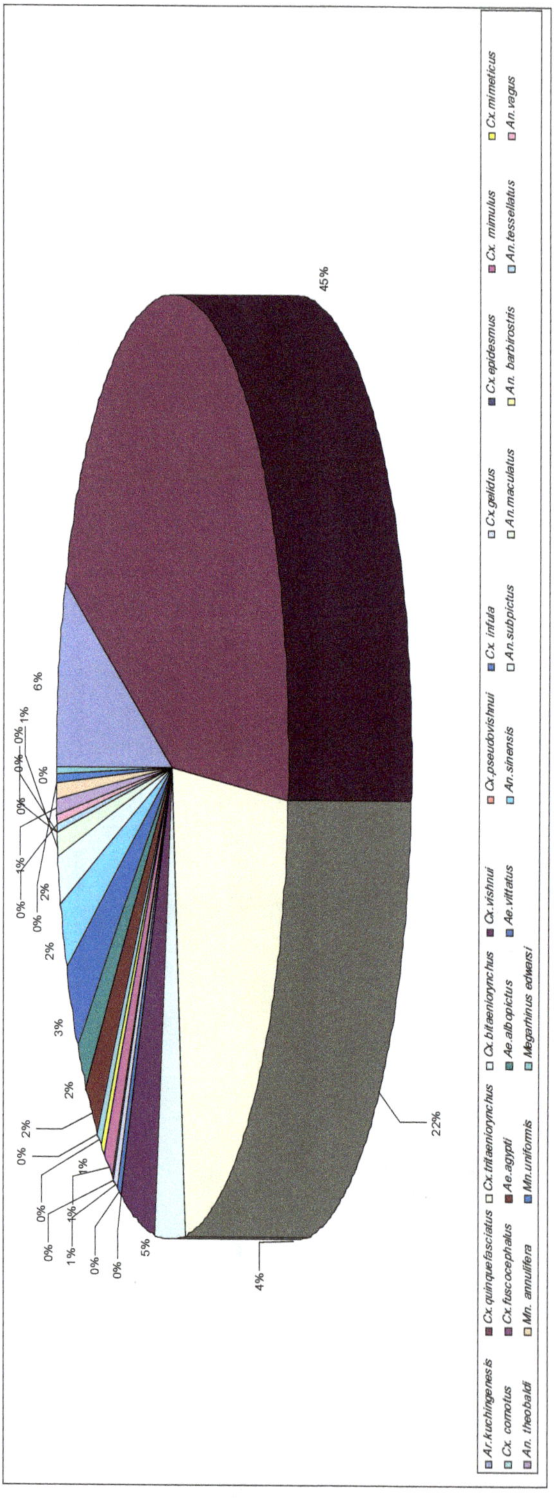

Figure 15.2: Density of different mosquito species recorded in Srivilliputhur Taluk during late post-monsoon season (March 2001 to May 2001).

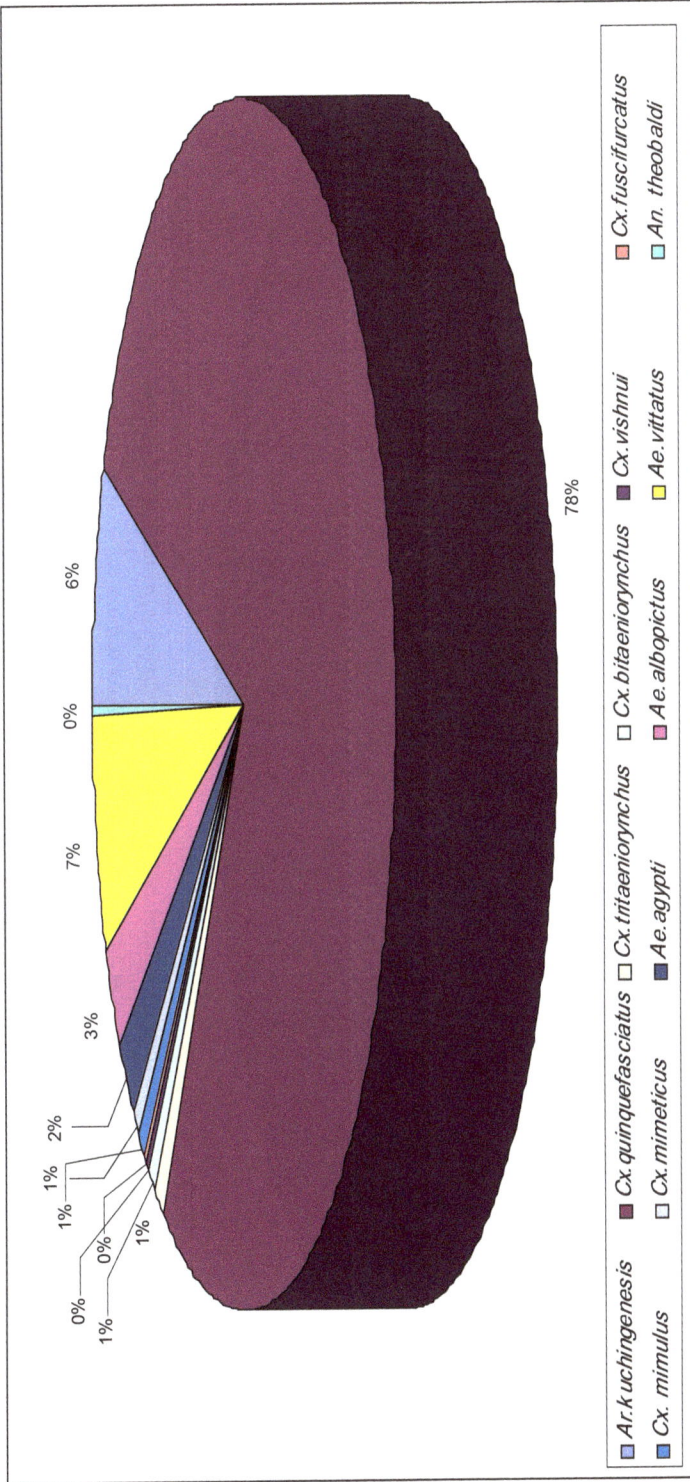

Figure 15.3: Density of different mosquito species recorded in Srivilliputhur Taluk during pre-monsoon season (June 2001 to August 2001).

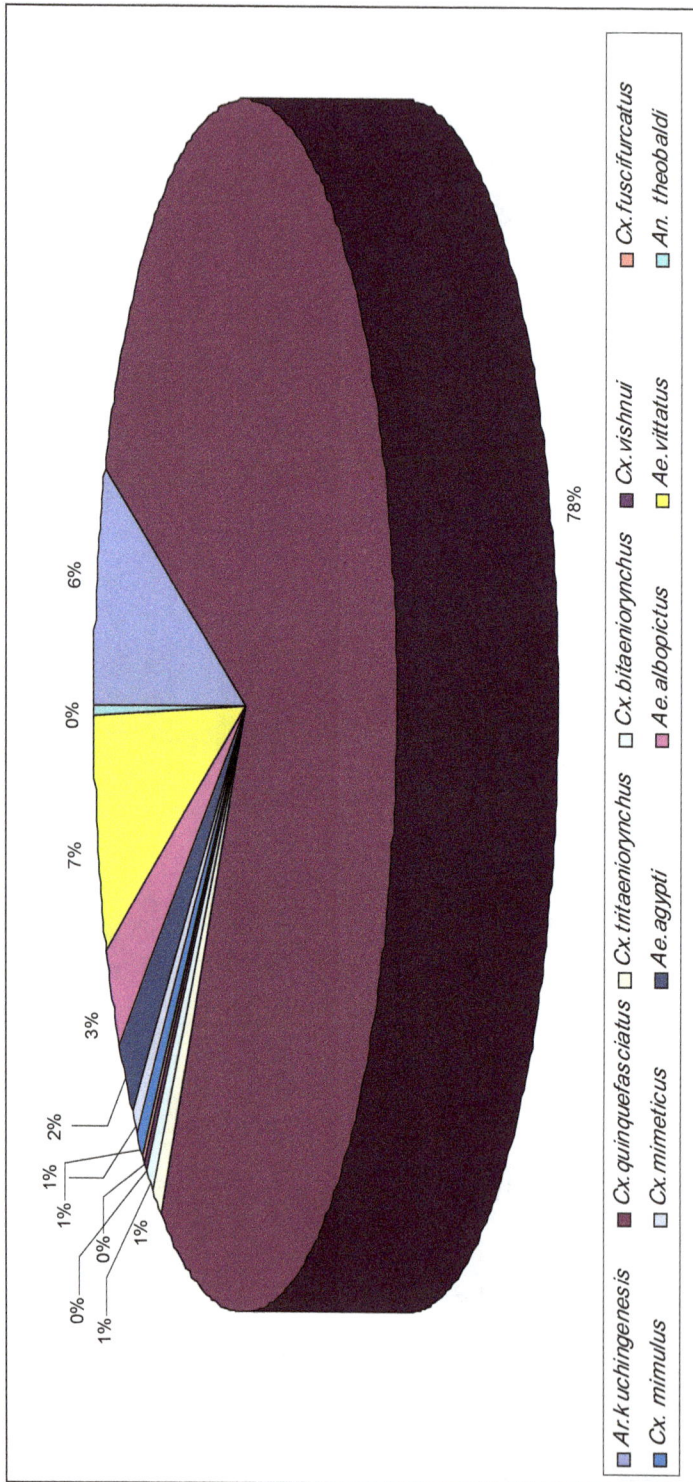

Figure 15.3: Density of different mosquito species recorded in Srivilliputhur Taluk during pre-monsoon season (June 2001 to August 2001).

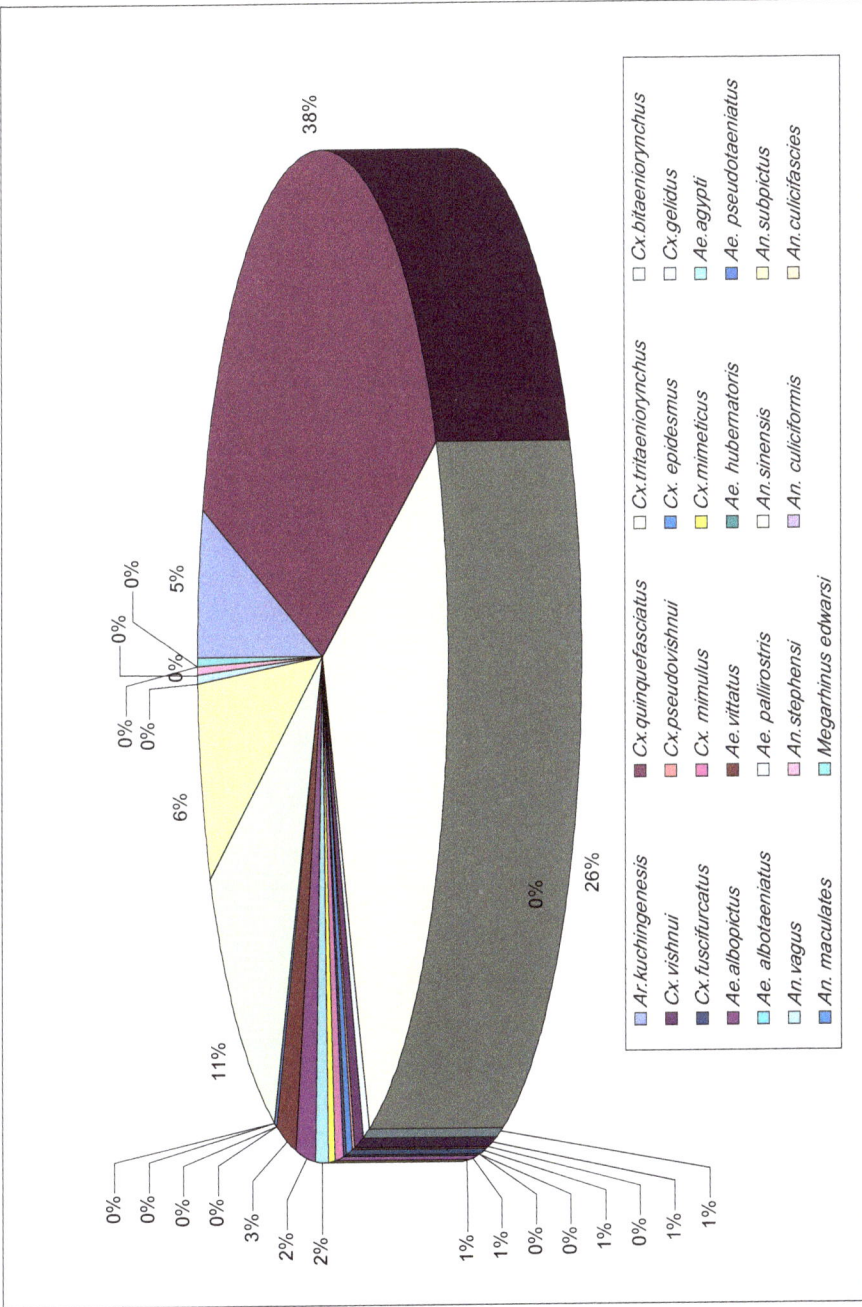

Figure 15.4: Density of different mosquito species recorded in Srivilliputhur Taluk during monsoon season (September 2).

and rural areas becomes the hosts for the mosquitoes. Pandian and Manoharan (1996) reported that high density of *Culex quinquefasciatus* was recorded in the central region of Madurai city than the periphery because of the presence of high sewage stagnation and thick human population. Other mosquito species such as, *Cx. tritaeniorhynchus, Cx. bitaeniorhynchus* and *Ar. kuchingenesis* showed moderate density. This may be due to the presence of agro-ecosystem which is a suitable breeding habitat for the first two species and stagnant sewage water for *Ar. kuchingenesis.* Similar results have been reported with the population of JE Vector mosquitoes in Uttar Pradesh (Kanojia *et al.,* 2003).

The remaining species showed less density throughout the study period. This may due to their breeding and feeding behaviours. Pandian *et al.* (1997) reported that the density of the mosquito species in a given area shall be influenced by the availability of suitable breeding habitats and feeding hosts, rain, irrigation for cultivation and water storage for human utilization etc affect the mosquito population. *Culex quinquefasciatus* showed highest density in all the four seasons. This may be due to the presence of breeding habitats throughout year. *Cx. tritaeniorhynchus, Cx. bitaeniorhynchus, An. subpictus* and *Ae. sinensis* density was high in all seasons except pre-monsoon. The population of this mosquito is associated with the cultivation of paddy. During the monsoon and post-monsoon seasons cultivation of paddy is high. These mosquito species coincided with the paddy cultivation and other irrigation system. Similar results were reported by Simsek (2004) in Turkey and Kanojia *et al.* (2003) in Uttar Pradesh. Other species of *Culex* and *Anopheles* genera showed variation in different seasons. The density of mosquito species showed variation in different seasons such as, early spring, late spring, summer and mid summer in Northern eastern United States (Cranes, 2004). The population *Cx. pipiens* and *Cx. perexiguus* increased following rainfall in East Africa (Chevalier *et al.,* 2004).

Table 15.3: Density and frequency of different mosquito species recorded in Srivilliputhur Taluk during pre-monsoon season (June 2001–August 2001).

Sl.No.	Name of the Species	Total	Density (per cent)	Frequency (per cent)
1.	*Ae. aegypti*	93	1.99	68.75
2.	*Ae. albopictus*	129	2.77	25
3.	*Ae. vittatus*	299	6.41	43.75
4.	*An. theobaldi*	82	0.18	12.5
5.	*Ar. kuchingenesis*	289	6.20	68.75
6.	*Cx. mimulus*	48	1.03	12.5
7.	*Cx. bitaeniorhynchus*	40	0.86	31.25
8.	*Cx. fuscifurcatus*	13	0.28	6.25
9.	*Cx. mimeticus*	40	0.86	12.5
10.	*Cx. quinquefasciatus*	3574	76.68	100
11.	*Cx. tritaeniorhynchus*	47	1.01	62.5
12.	*Cx. vishnui*	7	0.15	12.5
	Total	**4661**		

Table 15.4: Density and frequency of different mosquito species recorded in Srivilliputhur Taluk during monsoon season (Sep. 2001–Nov. 2001).

Sl.No.	Name of the Species	Total	Density (per cent)	Frequency (per cent)
1.	Ae. aegypti	225	1.71	93.75
2.	Ae. albopictus	328	2.49	43.75
3.	Ae. albotaeniatus	6	0.049	6.25
4.	Ae. hubernatoris	14	0.11	6.25
5.	Ae. pallirostris	7	0.053	6.25
6.	Ae. pseudotaeniatus	20	0.15	12.5
7.	Ae. vittatus	339	2.57	43.75
8.	An. culiciformis	10	0.08	6.25
9.	An. maculates	2	0.015	6.25
10.	An. culicifascies	4	0.018	6.25
11.	An. sinensis	1494	11.32	100
12.	An. stephensi	18	0.136	12.5
13.	An. subpictus	853	6.47	75
14.	An. vagus	27	0.205	25
15.	Ar. kuchingenesis	630	4.77	81.25
16.	Cx. epidesmus	131	0.99	25
17.	Cx. mimulus	133	1.01	12.5
18.	Cx. bitaeniorhynchus	84	0.64	50
19.	Cx. fuscifurcatus	17	0.13	6.25
20.	Cx.gelidus	12	0.09	18.75
21.	Cx. mimeticus	116	0.88	12.5
22.	Cx. pseudovishnui	10	0.08	6.25
23.	Cx. quinquefasciatus	5046	38.24	100
24.	Cx. tritaeniorhynchus	3470	26.30	87.5
25.	Cx. vishnui	163	1.24	75
26.	Mg. edwarsi	35	0.27	6.25
	Total	13194		

Among the four different seasons pre-monsoon season showed lowest density. This may be due to lack of breeding habitats. Temporary water bodies are barren and the permanent water sources also dried, which results in unfavourable condition for larval habitat. Since late post-monsoon was the entry of dry season, the density declined. But, during rainy and post-monsoon months, the water bodies in the present study sites filled and created favourable condition for oviposition of female and its successive larval development, hence, the density and number of species were more in these seasons. The mosquito population increased significantly following the raise in the level of the lake, following heavy rain fall (Robertson *et al.,* 1993). The heavy rainfall and flooding increased the mosquito breeding sites and the density of mosquito, resulted in the outbreak of arboreal disease in Australia (Kelly-Hope *et al.,* 2004; Hu *et al.,* 2006).

References

Chevalier, S., de la Rocque, S., Baldet, T., Vial, L. and Roger, F., 2004. Epidemiological processes involved in the emergence of vector-borne diseases: West Nile fever, Rift Valley fever, Japanese encephalitis and Crimean-Congo haemorrhgic fever. *Rev. Sci. Tech. Off. Int. Epiz.,* 23(2): 535–555.

Cranes, J.W., 2004, A classification system for mosquito life cycles: Life cycles types for mosquitoes of North eastern United States. *J. Vector. Ecol.,* 1: 1–10.

Kanojia, P.C., Shetty, P.S., and Geevarghese, G., 2003. A long-term study on vector abundance and seasonal prevalence in relation to the occurrence of Japanese encephalitis in Gorakhpur district, Uttar Pradesh. *Ind. J. Med. Res.,* 117: 104–110.

Kelly-Hope, L., Purdie, D. and Kay, B., 2004, Ross river virus disease in Australia, 1886–1998, with analysis of risk factors associated with outbreaks. *J. Med. Entomol.,* 41: 133–140.

Hu, W., Tong, S., Mengerson, K. and Oldenburg, B, 2006, Rainfall, mosquito density and the transmission of Ross river virus: A time-series forecasting model. *Ecological Modelling,* 196: 505–514.

Lopes, P., Lourenco, P., Sousa, C., Novo, T., Rodrigues, J., Almeida, A.P.G. and Seixas, J., 2005, Modelling patterns of mosquito density based on remote sensing images. In: *II International Conference and Exhibition on Geographic Information* (May 30th to June 2nd), Estoril Congress Centre.

Pandian, R.S. and Chandrasekaran, M.K., 1980, Rhythms in the biting behaviour of mosquito, *Armigeres subalbatus. Oecologia,* 47: 89–95.

Pandian, R.S. and Manoharan, A.C., 1994, Survey on the mosquitoes of Avaniapuram, a sewage fodder farm area of Madurai Corporation in Tamil Nadu, India. *Geobios, New Reports,* 13(2): 109–113.

Pandian, R.S. and Manoharan, A.C., 1996. Distribution pattern and biting behaviour of *Culex gelidus* Theobald: A *Japanese encephalitis* vector in Madurai. *J. Env. Pollu.,* 3 (2): 79–82.

Pandian, R.S., Vanithavalli, V., Tamilselvan, R. and Manoharan, A.C., 1997, Study on the bionomics of urban mosquitoes with reference to species diversity, spatial distribution pattern and preferential habitat selection. *Proceedings of the 2nd Symposium on Vector and Vector-borne Diseases,* p. 194–201.

Robertson, L.C., Prior, S., Aperson, C.S. and Irby, W.S., 1993, Bionomics of *Anopheles quadrimaculatus* and *Culex erraticus* (Diptera: Culicidae) in the Falls Lake basin. North Carolina: Seasonal changes in abundance and gonotrophic status and host-feeding patterns. *J. Med. Entomol.,* 30(4) : 684–698.

Simsek, F.M., 2004. Seasonal larval and adult population dynamics and breeding habitat diversity of *Culex theileri* Theobald (Diptera: Culicidae) in the Golbasi district, Ankara, Turkey. *Turk. J. Zool.,* 28: 334–337.

2013, Glimpses of Animal Biodiversity   *Pages* **158–163**

Editor: **Dr. K. Muthuchelian,** *Vice Chancellor, Periyar University, Salem*

Published by: **Daya Publishing House, NEW DELHI**

Chapter 16

Altitudinal Variation in Diversity of Mosquito in Thiruvannamalai Temple: A Hilly Area Srivilliputtur Taluk

K. Manikandan and S.P. Sevarkodiyone

Postgraduate and Research Department of Zoology,
Ayya Nadar Janaki Ammal College (Autonomous) Sivakasi – 626 12

ABSTRACT

Altitudinal variation of mosquitoes was studied by collecting the mosquitoes using human bait sampling method during the period of September 2009 to February 2010 from four habitats such as residential area, low altitude (78ft), transition (altitude) area (156ft) and high altitude (234ft) of an rural ecosystem at Venkateshwarapuram village near srivilliputtur. The study revealed that species richness and diversity of mosquitoes were higher in high altitude region (234ft) followed by residential site, transition (altitude) area (156ft) and low altitude (78ft). In high altitude region (234ft) the breeding ground was abundant and which were not frequency visited and used by human beings than the other three sites.

Keywords: *Diversity of mosquitoes, Altitudinal variation, Pattern of occurrence.*

Introduction

Mosquitoes are believed to have evolved around 170 million years ago during the Jurassic era (199-144 million years ago). The earliest known fossils was from the cretaceous era (144-65 million years ago). Mosquitoes are found throughout the world except in permanently frozen zones. Of the 3500 species (3/4) three quarter are native to humid tropics. Sub tropics and all the ambient bio-ecosystems are being enhanced the proliferation of mosquito species. The tremendous abundance of increase of mosquito population is promoted by unavoidable development of human activities. The prominent of them are the increase in human population, irrigation of crops, flooded rice fields, unplanned sanitation, urban development, the migration of bird, rainfall, temperature, humidity and dry arid condition and the changing life style of the population in cities, urban, sub-urban and rural areas of our nation (Devi and Jauhari, 2007). Seasonality and circardian rhythm of mosquito population as well as other ecological and behavioural features are strongly influenced by climatic factors such as temperature, rainfall humidity, wind and duration of day light. Both seasonal and daily activity patterns of mosquito vectors are required as baseline knowledge to understand the dynamics of vector-borne pathogens and have been widely studied for many mosquito species throughout the world (Loetti *et al.,* 2007).There are approximately 300 species of mosquitoes grouped into 41 genera. Human malaria is transmitters only by females of the genus *Anopheles.* Of approximately 430 species of *Anopheles* only 30-40 species transmit malaria (Ania, 2009). Biodiversity, distribution and density of mosquito population were studied in several areas. Mosquitoes are the vectors of various diseases such as malarial, dengue, yellowfever, encephalitis, filariasis, etc. To control the vector it is essential to know the diversity and density of mosquito species of a particular place (Sagandeep *et al.,* 1994). Hence the present study "Altitudinal variation in diversity and density of mosquitoes of Thiruvannamalai Temple, Srivilliputtur taluk" was undertaken with reference to seasonality and abiotic factors (Rainfall).

Study Site

The study was conducted in Thiruvannamalai (Hill) Temple and nearby residential area. It is encircled by agriculture fields and other kinds of mosquitogenic places and the height of the hill is approximately 234ft from the ground level. This temple is also called as "Then Thiruppathi".The number of visiting devotees from other districts of Tamil Nadu increased for many folds during the month of September. The study area also comprised of various types of human settlement and varying number of cattles and other animals that favour the mosquito population. Four sites have been selected on the basis of the location of breeding habitats and the availability of vertebrate hosts for the mosquitoes.

Methods

The biting mosquitoes were collected for a period of 12 hours continuously. Human-bait sampling method was used to study the diversity of mosquitoes adopted by Pandian and Chandrasekaran (1980). All the preserved wild–caught mosquitoes were identified upto species level by the Entomologists, Centre for Research in Medical Entomology (ICMR), Madurai. Mosquito diversity was evaluated using species richness index or alpha diversity (Southwood, 1978) to assess the degree of biodiversity. Knowledge on the pattern of occurrence of mosquitoes reveals the dimension of pattern of occurrence and the rate of existence in the selective study sites. This pattern of occurrence of mosquitoes was analysed by adopting the method of Rydzanicz and Lonc (2003).

Results

In this present investigation the diversity of mosquito fauna of Thiruvannamalai hill and nearby residential area with varying altitude have been observed and prevalence of mosquitoes throughout the period of study was analysed. The diversity of mosquitoes recorded in the study area includes eight species of five genera namely, *Aedes, Anopheles, Armigeres, Culex, Mansonia* (Table 16.1).

Table 16.1: Diversity of mosquito species recorded in the study area during the study period (September 2009–February 2010).

Sl. No	Name of the Species
1.	*Aedes* (*Stegomyia*) *aegypti* (Linnaeus)
2.	*Aedes* (*Aedimorphus*) *vittatus* (Bigot)
3.	*Anopheles* (*Cellia*) *stephensi* Liston
4.	*Anopheles* (*Cellia*) *subpictus* Grassi
5.	*Armigeres* (*Armigeres*) *subalbatus* (Coquillett)
6.	*Culex* (*Culex*) *pseudovishnui* Colless
7.	*Culex* (*Culex) quinquefasciatus* Say
8.	*Mansonia (Mansonioides) uniformis (Theobald)*

Table 16.2 Site-wise diversity of adult mosquito species collected from the study area during the study period (September 2009–February 2010).

Sl. No.	Name of the Species	Residential Site	Low Altitude 78ft	Transition (Altitude) Area 156ft	High Altitude 234ft	Total
1.	*Ae. aegypti*	+	–	–	+	2
2.	*Ae. vittatus*	–	–	+	+	2
3.	*An. stephensi*				+	1
4.	*An. subpictus*	+	+	+	–	3
5.	*Ar. subalbatus*	+	+	+	–	3
6.	*Cx. pseudovishnui*	+	+	+	+	4
7.	*Cx. quinquefasciatus*	+	+	+	+	4
8.	*Mansonia uniformis*	–	–	–	+	1
	Total	5	4	5	6	**20**

Diversity Index

Among the four sites of the study area, there was no marked difference in the diversity of mosquitoes except in high altitude region (234ft) (Table 16.2). The species diversity recorded from the study area was higher in acute rainy season with rich biodiversity index than the moderate rainy season and winter seasons. The biodiversity index (α) was calculated for all seasons. The biodiversity indices for the occurrence of mosquitoes in various sites of the three seasons namely acute rainy season (0.68), moderate rainy season (0.82) and winter season (0.83) respectively and this variation in the occurrence of the species as shown in the Table 16.3.

Table 16.3: Number of sites from which the mosquitoes recorded in study area during the study period (September 2009–February 2010).

Mosquito Species	Seasons								
	Acute Rainy Season		Total	Moderate Rainy Season		Total	Winter Season		Total
	Sep 2009	Oct 2009		Nov 2009	Dec 2009		Jan 2010	Feb 2010	
Aedes aegypti	–	1	1	2	2	4	2	1	3
Aedes vittatus	–	1	1	1	1	2	2	1	3
Anopheles stephensi	–	1	1	1	1	2	2	1	3
Anopheles subpictus	–	1	1	1	1	2	3	3	6
Armigeres subalbatus	–	1	1	1	2	5	3	2	5
Culex pseudovishnui	4	3	7	2	4	7	4	4	2
Culex quinquefasciatus	3	4	7	4	4	8	4	5	9
Mansonia uniformis	–	–	–	–	1	1	–	–	–
Occurrence	7	11	18	14	16	29	20	17	37
Biodiversity index (α)	0.62	0.75	0.68	0.83	0.83	0.82	0.84	0.80	0.83

Table 16.4: Diversity and species richness of mosquito recorded in the study area during the study period (September 2009–February 2010)

Name of the Species	Acute Rainy Season	Moderate Rainy Season	Winter Season
	Sep 2009-Oct 2009	(Nov 2009-Dec 2009)	(Jan 2010-Feb 2010)
Ae. aegypti	–	+	+
Ae. vittatus	+	+	+
An. stephensi	–	+	+
An. subpictus	+	+	+
Ar. subalbatus	+	+	+
Cx.pseudovishnui	+	+	+
Cx.quinquefasciatus	+	+	+
Mansonia uniformis	–	+	–
Total	5	8	7
Biodiversity index (α)	0.78	0.87	0.83

Species Richness

Occurrence of mosquitoes recorded from September 2009 to February 2010 in various sites of the study area showed that both the species richness and diversity index were higher in moderate rainy season Table 16.4.

Pattern of Occurrence

Pattern of occurrence of mosquitoes of the study area was also calculated. Among the eight species recorded, *Culex quinquefasciatus* and *Culex pseudovishnui* were the two species exhibited constant pattern, where as the remaining six species were recorded as infrequent species Table 16.5.

Table 16.5: The pattern of occurrence of mosquito species collected in the study area during the study period (September 2009–February 2010).

Name of the Species	Diversity Pattern of Mosquitoes	Percentage	Total Number of Mosquito Species
Culex quinquefasciatus	Constant	100	**2**
Culex pseudovishnui	Constant	91.6	
Anopheles subpictus	Infrequent	37.50	**6**
Aedes aegypti	Infrequent	33.33	
Armigeres subalbatus	Infrequent	33.33	
Anopheles stephensi	Infrequent	25	
Aedes vittatus	Infrequent	25	
Mansonia uniformis	Infrequent	4.16	
Total			**8**

Discussion

The result of this study indicates that moderate rainy season appeared to have higher biodiversity index than the remaining seasons. This was due to the continuous precipitation that was followed by occurrence of breeding habitats. From the results obtained, both November-2009 and December (Moderate rainy season) exhibited higher prevalence of intergeneric mosquito diversity. This was due to the existence of varieties of water bodies following the rain and the existence of different vegetation cover.

Altitudinal Variation

During the period of study the diversity and density of mosquito species differ from different altitudes. The high altitude region (234ft) shows high species richness which includes 6 species namely *Aedes aegypti, Aedes vittatus, Anopheles stephensi, Culex pseudovishnui, Culex quinquefasciatus* and *Monsonia uniformis* were found. Because the high altitude site has varieties of breeding grounds which were not reached by human. That may be the reason for higher diversity pattern of mosquitoes.

References

Ania.S.A., Bajo, A.D and Jonarthan, K., 2009. Efficacy of some plant extracts on *Anopheles gambiae* mosquito larva, Nigiria. *Academic Journal of Entomol.*, 2(1): 31–35.

Devi, N.P. and Jauhari, R.K., 2007. Mosquito species associated within some Western Himalayas Phytogeographic Zones in the Garhwal region of India. *J. Ins. Sci.*, 7: 33–37.

Loetti, V., Burroni, N. and Vezzani, D., 2007. Seasonal and daily activity patterns of human biting mosquitoes in a wetland system in Argentina. *J. Vect. Eco.*, 32(2): 358–365.

Pandian, R.S. and Chandrashekaran, M.K., 1980. Rhythms in the biting behaviour of mosquito *Armigeres subalbatus. Oecologia* (Berl.), 47: 89–95.

Rydzanicz, K. and Lonc, E., 2003. Species composition and seasonal dynamics of mosquito larvae in the Wroclaw Poland area. *J. Vec. Ecol.*, 23(2): 255–266.

Sagandeep, V.C. Kappeor and Grewal, J.S., 1994. Some mosquito (Diptera: culicidae) species of Punjab and Himachal Pradesh. *J. Insc. Sci.*, 7(1): 48–50.

Southwood, T.R.F., 1978. *Ecological Method with Particular Reference to the Study of Insect Population,* 2nd edn. The ELBS. University Printing, Oxford Pub., U.K., p. 420–421.

2013, Glimpses of Animal Biodiversity　　　　　*Pages* **164–172**

Editor: **Dr. K. Muthuchelian,** *Vice Chancellor, Periyar University, Salem*

Published by: **Daya Publishing House, NEW DELHI**

Chapter 17

Biodiversity of Earthworm Resources and their Relation to Soil Quality in Mamsapuram Village, Srivilliputtur Taluk, Tamil Nadu

V. Mariappan, L. Isaiarasu and A. Marikani

Post-graduate and Research Department of Zoology
Ayya Nadar Janaki Ammal College (Autonomous),
Sivakasi – 626 124, Tamil Nadu

ABSTRACT

The present study was carried out during September, 2009 to record the biodiversity of earthworms in terms of their distribution, density, biomass and to investigate the relationship between earthworm abundance and soil quality in ten different pedoecosystem which comprises five from cultivated land (Paddy field, Sugarcane field, Banana plantation, Coconut grove and Mango farm) and other five from non-cultivated land (Fallow land, Grass land, Municipal waste dumped site, Sewage site and Industrial waste dumped site) around Mamsapuram village,(Srivilliputtur Taluk) which is 8 km away from Rajapalayam Town, bordering Western Ghats, one of the twelve biodiversity hotspots in the world were reported. A total of five species of earthworms belonging to four different families *viz.*, Megascolecidae, Glossoscolecidae, Eudrilidae and Octochaetidae and these species namely: *Lampito mauritti, Perionyx excavatus, Pontoscolex corethrurus, Eudrilus eugeniae* and *Dichogaster bolaui* were noted. The species differed in relation abundance in each of the habitats sampled, among the cultivated land, the total number of earthworms ranged from 40/m^2

to 89/m² in mango farm and banana plantation respectively. In non-cultivated land, the earthworm's number varied from 23/m² to 57/m² in fallow land and grassland respectively. The relative density (RD, per cent), relative biomass (RB, per cent), relative frequency (RF, per cent), species richness (D), Shannon-Weiner index (H) and evenness index (E) were also calculated. Data on the abundance of earthworms were correlated with soil physico-chemical parameters such as temperature (T), pH, electrical conductivity (EC), moisture (M), organic matter (OM), organic carbon (OC), total nitrogen (TN), total phosphate (TP), total potassium (TK), and C:N ratio. The analysis showed that earthworm abundance found in all ten different pedoecosystem was positively correlated with all physico-chemical parameters except temperature and electrical conductivity. In an other words, the earthworm abundances were negatively correlated with those two factors.

Keywords: *Earthworms, Relative density, Relative biomass, Relative frequency, Species richness, Pedoecosystem.*

Introduction

Soil is home to several kinds of animals, and among these earthworms form a major component. They are known to inhabit the earth for the last 600 million years, and silently working day and night in improving the quality of the soil (Julka, 2008). They play a prominent role in regulating soil processes and known to improve soil productivity and fertility by enhancing physical, chemical and biological characteristics of soil (Lee, 1985). Hence, they have been recognized as the most important soil ecosystem engineers (Lavelle *et al.,* 1997; Santra and Bhowmik, 2001).

India is one of the very rich country as far as its biodiversity concern particularly in terms of biological diversity due to its unique bio-geography location, diversified climatic conditions and enormous eco-diversity and geo-diversity (Venkataraman, 2009). As many as 418 species of earthworms are known from our country, which harbors about 11.1 per cent of the global earthworm diversity, estimated at 4000 species (Julka, 2008). Totally 402 species and sub species of earthworm belonging to 67 genera and 10 families have been identified from our country (Tripathi and Bhardwaj, 2004). Abundance the diversity of earthworm population in the soil are normally correlated to their a biotic factors such as soil, moisture, temperature, humidity, altitude and agricultural practices (Handricksen, 1990). Thus, their distribution are governed by several ecological and edaphic factors (Tripathi and Bhardwaj, 2004).

Materials and Methods

Study Area

An extensive survey of earthworms was carried out in and around Mamsapuram village, which is 8 km away from Rajapalayam town (10 29 N; 79 110 MSL and 54 E) located in the foothills western ghats, which is one of the 12 biodiversity hot spot in the world were reported.

Sampling

There are 10 different sampling sites were chosen for the present survey of earthworms at the sampling sites mainly includes 5 different cultivated lands, which mainly comprises paddy field, sugarcane field, banana plantation, coconut groove and mango farm and other 5 different sites from non-cultivated lands, namely fallow land, grassland, municipal garbage dumped site, sewage site and industrial waste damaged site. The earthworm survey of present study was carried out in the month of September 2009, and the survey was done in the middle of the month.

Collection of Earthworm

Earthworms were collected from different habitats by diggings the soil and hand sorted (Julka and Paliwal, 1993). The grasses and herbs were first cleared from the surface of the sampling sites and a plot of total of 8 quadrates of 25 cm × 25 cm × 20 cm per sampling sites per sampling occasion. When the soil was drug out and earthworms were hand shorted and then they were carefully transported to the laboratory for further process sing with in their native soil by west cloth bags. In the lab, collected earthworms were identified by using of simple key given by Julka (2008) and Tripathi Bhardwaj (2005).

Study of Earthworms

Density of earthworms was calculated as the number of individual (both clitellates and non clitellates) (individual/m^2). Biomass of worms was determined in on electronic balance with 0.01 mg accuracy and values are given in fresh weight basis (with gut content) (g/m^2). Relative density (RD) relative frequency (RF) and relative biomass (RB) (Dash, 1993) species richness (Margalef, 1968), Diversity index (Shannon and Weiner, 1963), index of evenness (Pielou, 1966) were calculated.

Soil Analysis

Soil samples were taken from each site for analysis. Temperature at 10 cm depth was measured by soil thermo meter and pH (1.5w/v) was determined by using pH meter. Soil moisture content was estimated gravimetrically on a wet weight basis by oven drying at 105 (Gupta, 2000) and electrical conducting was measured by using conductivity meter (Jackson, 1973). Available organic carbon (OC), organic matter, total nitrate, total phosphate, total potassium were analyzed by Sathasivam and Manikam (1992), Gupta (2000) Subbiah and Asija (1956) Olsen and Wetanoble (1985) and Standford and Friesdsh (1949) respectively. C:N ratio and simple correlation were also calculated.

Results and Discussion

Earthworm diversity is much more in natural system than in interfere habitat (Lee 1985). They are found in all types of soil provided there is sufficient moisture and food supply (Ghosh, 1993). They are fundamental to the dynamics of in ecosystem. Both physical and organic factors of soil have been known to influence the abundance and distribution of earthworms (White, 1975). Sathianarayanan and Khan (2006) were reported that a total of 10 species of earthworms include *D. Willisi, D. limella, D. scendens, Pontodrilus bermindensis, P. Corethrurus, L. mauritti, P. excavatus, E. eugeniae, Octachaetona, Serrata* and *Obarnesi,* belonging to 7 genera and 6 families noted in 14 localities from different habitats of pondicherry region. Karmegam and Daniel (2007), were reported a total of 10 different earthworms belonging to 7 genera and 6 families noted in 14 localities from different habitats of Pondicherry region.

A total of 5 species belonging to 4 different families of earthworms were recorded from study areas *i.e.* in and around Mamsapuam village, Srivilliputtur taluk, Tamil Nadu (Table 17.1). The identified species comprised of *P.corethrurus* (Muller), *P. excavates* (Perrier), *E. eugeniae* (Kinberg) and *D. bolani* (Michaelsen) an families were Megascolecidae Glosioscolecidae, Eundridae and Octochaetidae respectively (Tables 17.1 and 17.2). Karmegam and Daniel (2007), were reported a total of ten earthworm species belonging to four families, namely, Megascolecidae (three species) Octochaectidae (three species) Moniligastridae (three species) and Glossoscolecidae (one species), from different location at ten sites in Dindugul district, Tamil Nadu, South India, during 1997-1999. Earthworm density and biomass fluctuated seasonally. Density was relatively constant (70-100m^2) during the wet season (November –May) and lower (20-30m^2) in the drier months (August–October). Biomass increased rapidly over the wet late winter to early spring months (February-April) but decreased to <10 gm^2 by October. The distribution of organic matter in soil influence the distribution greatly and soils that are poor organic matter do not usually support large number of earthworms (Edwards and Bohlen, 1996; Kale, 1998). Out of 5 speceis, *L.marutti* was present in seven sites, namely, PF, SCF, BP, MF, G2, MWDS, SS, and absent in three sites in CG, FL and IWDS and absent in four sites, PF, SCF, FL and SS, *P. excavatus* was present in eight sites, PF, SCF, BP, CG, MF, GL, SS, IWDS and absent in two sites, FL and MWAS. *E.eugeniae* was present in eight sites PF, SCF, BP, CG, FL, GL, MWDS, IWDS and absent in MF and SS. *D. bolani* was present in six sites PF, SCF, CG, MF, FL, SS and absent in four sites namely, BP, GL, MWDS and IWDS (Tables 17.1 and 17.2).

The overall higher number of earthworms were recorded in the banana plantations of the cultivated land, which was found to be 33 individuals of *P. excavatus* and its RD WAS 37.08 per cent and overall least number was recorded in mango farm of cultivated land of the same species, which was 8 individuals and its RD was 20 per cent (Table 17.1). In contrast, out of five different sites of non-cultivated land, the overall highest number of species of *P.excavatus* was found to be 23 individuals and its RD was 40.35 per cent in grassland whereas, overall smallest number was present in an industrial waste dumped site and earthworm species *P.corethrurus* and its value was 6 individuals and its RD was 21.43 per cent (Tables 17.1 and 17.2). Interestingly, among the various pedoecosystem studied during September, 2009, in the cultivated and non-cultivated land, highest total number of earthworms were reported in banana plantation and grass land, and lowest number was found in mango farm and fallow land respectively (Table 17.3).

Species diversity index (H') of the earthworm in 10 different pedoecosystem of Mumsapuram area were also studied. Among this, paddy field of the cultivated land showed the highest value, 1.38 and least value, 0.68 observed in the fallow land of non cultivated land. Stable ecosystem have high species diversity as compared to unstable environments (May, 1997). Senapathi and Dash (1981) recorded lower index of species

Table 17.1: Showing population structure of Earthworm found in cultivated land of different pedoecosystem in Mamsapuram village during September 2009

Name of the Pedoecosystem	Sample Identification	Earthworm Species			Population Structure					
			Biomass g/m²	Density no/m²	RD per cent	RF per cent	RB per cent	Evenness Index	Spp. Richness	Diversity Index
Paddy Field	PF	L.mauritii	18.1	20	23.26	18.87	22.56	0.993	0.67	1.38
		P.excavatus	14.04	20	23.26	14.18	17.5			
		E.eugenia	31.22	29	33.72	19.6	38.91			
		D.bolaui	16.87	17	19.76	20	21.03			
Sugarcane Field	SF	L.mauritii	19.82	26	33.77	24.53	26.96	0.986	0.69	1.37
		P.excavatus	14.2	16	20.78	11.35	19.32			
		E.eugenia	26.02	20	25.97	13.51	35.4			
		D.bolaui	13.47	15	19.48	17.65	18.32			
Banana Plantation	BP	L.mauritii	16.78	22	24.72	20.75	22.89	0.971	0.67	1.35
		P.corthrurus	16.22	18	20.22	21.69	22.12			
		P.excavatus	22.91	33	37.09	23.4	31.24			
		E.eugenia	17.42	16	17.98	10.81	23.76			
Coconut Groove	CG	P.corthrurus	17.29	20	23.53	24.1	22.12	0.942	0.68	1.31
		P.excavatus	19.41	26	30.59	18.44	24.83			
		E.eugenia	34.26	30	35.29	20.27	43.83			
		D.bolaui	7.21	9	10.59	10.59	9.22			
Mango Farm	MF	P.corthrurus	9.21	11	27.5	13.25	27.25	0.936	0.54	1.02
		P.excavatus	6.78	8	20	5.67	20.06			
		D.bolaui	17.81	21	52.5	24.7	52.7			

Table 17.2: Showing population structure of Earthworm found in non cultivated land of different pedoecosystem in Mamsapuram village during September, 2009

Name of the Pedoecosystem	Sample Identification	Earthworm Species	Population Structure							
			Biomass g/m²	Density no/m²	RD per cent	RF per cent	RB per cent	Evenness Index	Spp. Richness	Diversity Index
Fallow Land	FL	E.eugenia	12.11	10	45.45	5.76	51.86	0.986	0.32	0.68
		D.bolaui	11.24	13	59.09	15.29	48.14			
Grass Land	GL	L.mauritii	4.29	8	14.03	7.55	11.47	0.950	0.74	1.32
		P.corthrurus	6.78	12	21.05	14.46	18.12			
		P.excavatus	14.57	23	40.35	16.31	39.21			
		D.bolaui	11.41	14	24.56	9.46	30.5			
Municipal Waste Dumped Site	MWDS	L.mauritii	7.81	12	28.57	11.32	26.21	1.009	0.53	1.10
		P.excavatus	9.71	16	38.10	19.28	32.58			
		E.eugenia	12.28	14	33.33	9.46	41.21			
Sewage Site	SS	L.mauritii	13.71	18	50	16.98	51.91	0.954	0.56	1.04
		P.excavatus	6.38	8	22.22	5.67	24.16			
		D.bolaui	6.32	10	27.78	11.76	23.93			
Industrial Waste Dumped Soil	IWDS	P.corthrurus	4.84	6	21.43	7.23	20.7	0.927	0.60	1.01
		P.excavatus	5.27	7	25	4.96	22.53			
		E.eugenia	13.28	15	53.57	10.14	56.78			

Table 17.3: Showing contribution of various physico-chemical parameters to earthworms collected at different pedoecosystem of Mamsapuram village, during September, 2009.

Name of the Pedoecosystem	Sample Identification	Physico-chemical Parameters									
		Tem. (°C)	Moisture (%)	pH	OM (%)	OC (%)	TN (%)	TP (%)	TK (%)	EC µmhos/cm	C:N Ratio
I. Cultivated land											
Paddy Field	PF	26.4	26.09	8.19	24.82	14.40	1.34	0.42	0.82	0.79	10.75
Sugarcane Field	SCF	28.2	24.88	8.10	27.41	16.20	1.70	0.38	0.76	0.80	9.53
Banana Plantation	BP	27.6	25.12	8.24	24.41	12.42	1.22	0.32	0.66	0.72	10.18
Coconut Grove	CG	26.3	24.42	8.20	24.48	14.20	1.24	0.37	0.72	0.50	11.45
Mango Farm	MF	28.1	23.56	7.96	19.68	11.42	0.76	0.30	0.71	0.84	15.02
II. Noncultivated land											
Fallow Land	FL	28.4	19.42	7.89	15.17	8.80	0.86	0.28	0.62	1.23	10.23
Grass Land	GL	26.7	25.38	7.67	16.76	9.72	0.78	0.30	0.68	1.28	12.46
Municipal Waste Dumped Site	MWDS	27.3	21.32	7.74	17.65	10.24	0.91	0.32	0.63	1.48	11.25
Sewage Site	SS	28.0	22.08	7.40	19.17	11.12	0.95	0.25	0.65	1.38	11.70
Industrial Waste Dumped Site	IWDS	28.2	20.28	7.76	16.77	9.73	0.72	0.27	0.63	1.32	13.51

dominance for high distributed land grazed pasture. The presences of a species in a particular habitat and its absence from other habitats shows the species specific distribution of earthworms in different pedoecosystem. Similar conclusion have been observed from species composition of earthworms in different grassland, cultivated and forest soil (Singh, 1997).

In terns of number of species reported in the various pedoecosystem studied, four species were found in PF, SCF, BP, CG, and GL, three species were found in MF, MWDS, SS, and IWDS, and two species were observed in fallow land the occurrence of different species of earthworms were studied in relation to physico-chemical properties of there soil habitats (Table 17.3). In the month of September 2009, total number of earthworms in all the 10 different study site were positively correlated with pH, moisture, OC, OM, TN, TP, TK, and C:N ratio, were as negatively correlated with temperature and EC (Table 17.3). Maximum index of species richness (d) with a value of 0.74 in grass land and lowest value of 0.32 were observed in fallow land and both these habitats comes under non cultivated land. In contrast, its value was high in sugarcane field (0.69) and the low value was observed in mango farm of cultivated land. Maximum index of species evenness (e = 1.009) was estimated in municipal waste dumped soil and lowest value was recorded in an industrial waste dumped site of non cultivated land.

References

Dash, M.C., 1993. *Fundamentals of Ecology*. Tata McGraw-Hill Publishing Company Ltd., New Delhi.

Edwards, C.A. and Bohelen, P.J., 1996. *Biology and Ecology of Earthworm*. Chapman and Hall, Londan.

Ghosh, A.H., 1993. *Earthworm Resources and Vermiculture*. Zoological Survey of India, Calcutta.

Gupta, P.K., 2000. *Methods in Environmental Analysis Water, Soil and Air*. Agrobios, Jodhpur.

Handricksen, N.B., 1990. Leaf litter selection by detritivores and geophagous earthworm, *Biol. Fertil. Soils*, 10: 17–21.

Jackson, M.I., 1973. *Chemical Analysis*. Practice Hall of India Pvt. Ltd., New Delhi, p. 180–185.

Julka, J., 2008. *Know your Earthworms*. Foundation for Life Sciences and Business Management, Solan, H.P.

Julka, J.M., and Paliwal, R., 1993. Collection, preservation and study of earthworm. In: *Earthworms Resources and Vermiculture*, (Ed.) J.M Julka. Zoological Survey of India, Calculta, p. –11.

Kale, R.D., 1998. Earthworm nature's gift for utilization of organic wastes. In: *Earthworm Ecology*, (Ed.) C.A. Edwards. Soil and Water Conservation Society Ankeny, Lowa, p. 355–376.

Karmegam, G.M. and Daniel, M., 2000. The disposal of soil waste by utilization of vermitechnology. *J. Eco. Biol.*, 17: 257–272.

Lee, K.E., 1985. *Earthworms: Their Ecology and Reationships with Soil and Land Use*. Acadamic Press, London, 411 p.

Margalef, R., 1968. *Perspectives in Ecological Theory*. University of Chicago press, Chicago, p. 112.

May, R.M., 1979. The structure and dynamics of ecological communities. In: *Population Dynamics: 20ᵗʰ Symposium of British Ecological Society,* (Eds.) R.M. Anderson, B.D. Turner and L.R. Taylar. Blackwell, Oxford, pp. 385–407.

Olsen, S.R. and Watanoble, F.S., 1985. Test of Ascorbic acid method for determining phosphorus in water and $NaHCO_3$ extract for soil. *Soil. Am. Proc.*, 29: 600–663.

Paravaresh, A., Muvahedian, K. and Hamidian, L., 2004. Vermi stabilization of municipal waste water sludge with *Eisenia fetida. Iranian J. Env. Health Sci. Eng.*, 1(2): 43–50.

Pielon, E.G., 1966. The measurement of diversity in different types of biological collection. *Journal of Theoretical Biology,* 13: 45–163.

Sadasivam, S. and Manickam, A., 1992. *Biochemical Methods for Agricultural Sciences*. Wiley Eastern Limited, New Delhi, 34 p.

Santra, S.K. and Bhowmik, K.L., 2001. Vermiculture and development of agriculture. *Yogina,* p. 43–45.

Sathianarayanan, A. Akisa and Khan, B., 2006. Diversity, distribution and abundance of earthworm in Pondichery region. *Tropical Ecology,* 47: 139–144.

Senapati, B.K. and Dash, M.C., 1981. Effect of grazing on the elements of production in the vegetation and Oligochaete components of a tropical pasture land. *Review on Ecology and Biology of Soils*, 18: 487–505.

Shannon, C.E. and Weiner, W., 1963. *The Mathematical Theory of Communication*. University of Illiusis Press, Urbana, p. 117.

Singh, J., 1997. Habitat preference of selected Indian earthworm species and their efficiency in reduction of organic material. *Soil. Biol. Biochem.*, 29(3/4): 585–588.

Standford, S. and Friedsh, I., 1949. Use of frame photometer in rapid soil tests of K and Calcium. *Agram. J.*, 41: 446–447.

Subbiah, B.V. and Asija, C.S., 1956. A prepaid procedure for estimation of available Nitrogen in soil. *Curr. Sci.*, 25: 259–260.

Tripathi, G. and Bhardwaj, P., 2004. Earthworm diversity and habitat preference in arid Region of Rajasthan. *Zoo Print Journal*, 19(17): 1515–1519.

Venkataraman, K., 2009. Biodiversity conversation in India. In: *Proceeding of Conference on Resent Advances in Applied Zoology*, ANJAC College, p. 99–108.

White, W. (Jr.), 1975). *An Earthworm is Born*. Starting Publishing Company, New York.

2013, Glimpses of Animal Biodiversity   *Pages* **173–185**
Editor: **Dr. K. Muthuchelian,** *Vice Chancellor, Periyar University, Salem*
Published by: **Daya Publishing House, NEW DELHI**

Chapter 18

Consequences of Climatic Change for the Zooplankton Biodiversity (Copepods) from Lake Velachery, Chennai, India

*K. Rajabunizal and R. Ramanibai**

Unit of Biomonitoring, Department of Zoology,
University of Madras, Guindy campus, Chennai

ABSTRACT

Zooplankton play an important role in the ecosystem as a critical intermediary in the food chain converting algae into invertebrate tissue. The water temperature of Velachery lake ranged from a minimum of 19°C in June to a maximum of 28°C in May, 2009. The air temperature ranged from 29°C in December to 39°C in May, 2009. The Climatic change has more impact on the Copepods in Velachery lake by producing abnormal exotopic protrusions or Tumor-like abnormalities (TLAs) and Protozoan ciliate epibiont's infestation on the freshwater species.

These Tumor–like abnormalities (TLAs) have been identified as a serious emerging threat to the food web in lakes. At certain times these protrusions can affect 27–54 per cent of the copepods of certain species. Cyclopoid copepods like *Mesocyclops aspericornis, M. pehpeiensis, M. thermocyclopoides, M. ogunnus, Eucyclops serrulatus* appear to be vulnerable to infestation. Protrusions are seen more frequently on adult females and nauplii than copepodites and males.

---

* Corresponding Author: E-mail: rramani8@hotmail.com

The freshwater copepods are heavily infested by protozoan ciliate epibionts such as *Epistylis niagarae, Vorticella* sp., which belong to the class *Peritrichia*.It was observed that the females were more susceptible to epibiont infestation than the males and the copepodites. The percentage frequency of infestation was found to be 53 per cent in females, 31 per cent in males and 16 per cent in copepodites. From our observation it was clearly noticed that the prosome region of copepods were highly infested (57 per cent) than the urosome region (43 per cent). None of the copepod nauplii were infested needs further study.

Since Lake Velachery is a eutrophicated one, it is necessary to identify hot spots that would give insight into possible causes for the prevalence of protrusions and protozoan ciliate infestation in zooplankton population. The appearance of protrusions in recent years may indicate the emergence of a global phenomenon with a common cause.

Keywords*: Epistylis niagarae, Vorticella sp., Exotopic protrusion, Eutrophication, Tumor.*

Introduction

Ecologically copepods are the link between primary producers (phytoplankton) and higher tropic levels (fish, birds). They are the ubiquitous members of the planktonic fauna. Because of these protrusions, copepod species are considered a serious threat to the food web in lake Velachery, Chennai, India. Tumor-like abnormalities were found on *Limnocalanus macrurus* in Lake Michigan (Vanderploeg *et al.,* 1998). The frequency of protrusions among the Great Lakes region ranged from 0 per cent to 70 per cent (Omair *et al.,* 1999). Crisafi and Crescenti (1975 and 1977) found exophytic lesions in 40 species of copepods from the Mediterranean sea. Silina and Khudolei (1994) reported exotopic protrusions on copepods from the Gulf of Finland. The histology of protrusions showing necrosis on copepods from Lake Michigan had been observed by Omair *et al.* (2001). The histological description of protrusions bears a striking resemblance to cysts of an ellobiopsid parasite on copepods in Lake Michigan (Bridgeman *et al.,* 2000).

Several ciliate protozoans live as epibionts on animals and plants using them as substrate (Sleigh 1979; Dragesco 1986; Fenchel 1987). The body surface of Crustaceans, including copepods serve as a convenient living environment for many groups of organisms. These include the so called epibionts (protozoa, bacteria, algae and various invertebrates) which settle on the body surface of other living organisms that there by become basibionts (Fernandaz–Leborans and Tato–Porto 2000) Epibiosis is advantages for the epibiont's dispersion and geographical expansion (Wahl 1989). High densities of epibiontic ciliates decrease copepods' ability to move (Overstreet 1983; Gortz 1996). It can also influence the growth and moulting process.

The aim of the current study was to describe the phenomenon of epibiosis and the presence of the epibiont *E. niagarae* on *M. aspericornis*.

Materials and Methods

Lake Velachery is one of the prime lakes situated in the urbanized area of the metropolitan city of Chennai in India. Lake is located at 12° 59' 15" Latitude and 80°

30' 45" Longitude. It is a shallow water body with a surface area of 265.48 acres and a maximum depth of 3.8 m (Figure 18.1). Zooplankton was sampled by duplicate tows of 6 cm diameter using 150 μm mesh net towed vertically from 1m off the bottom to the surface. Plankton specimens were collected from Velachery lake during Sep. to Nov. 2009 from 4 locations identified for the present study. Immediately after collection they were transported within half an hour and preserved in 4 per cent formalin (Edmondson 1959). Two or three 1 ml aliquots from each sample were examined using a stereoscope binocular microscope (Nikkon). For identification of *Epistylis niagarae* and *Mesocyclops aspericornis* Edmondson (1959), Cheng (1964) were used. All adult copepods were identified up to species level and immature copepodites up to genus, while nauplii were combined under one single group. All copepods with the epibionts were sorted, counted and studied individually. Morphometric measurements were calculated using an ocular micrometer.

Copepods with the protuberances were isolated, counted and examined individually. Histological analysis was performed for the abnormal protrusions (Howard and Smith, 1983). The tissue sections were observed at 400 ~ 1000 X with a stereoscope binocular microscope (Nikkon).

Results

Totally 5 species of cyclops were exhibited more number of protrusions on their body. The percentage of incidence shown by *Mesocyclops aspericornis* was 54 per cent, *Mesocyclops pehpeiensis* 42 per cent, *Mesocyclops ogunnus* 40 per cent, *Mesocyclops thermocyclopoides* was 36 per cent and *Encyclops serrulatus* 27 per cent (Figure 18.2).

Based on the microscopical observation it was observed that out of 422 ind/l of *M. aspericornis* isolated, only 325 ind/l were infested with the epibiont *E. niagarae* and the remaining 97 ind/l were non-infested. The percentage of infestation estimated revealed that the females were more susceptible towards epibiont's infestation (53 per cent) than the males and the copepodites (31 per cent and 16 per cent). To our surprise, nauplii encountered during the study were not infested by these epibionts. The reason for this is yet to explore. Analyses of samples collected during the month of Nov. 2009 showed a maximum percentage (23 per cent) of infestation on the prosome of *M. aspericornis* and minimum percentage (15 per cent) of infestation during Oct. 2009 (Table 18.1).

During the sampling period the percentage of infestation 57 per cent occurred on the prosome of *M. aspericornis* was found to be more than the urosome 43 per cent (Table 18.1). It was always noticed that the epibionts were found to be attached to the body surface, antennae, egg sac and on the setae (Figures 18.11– 18.15).

Taxonomic Position of *E. niagarae* (Sprague and Couch 1971) (Figure 18.16)

Phylum   :   Protozoa (Honigberg 1964).

Sub Phylum   :   Ciliophora (Doflein 1897).

Class	:	Oligohymenophorea (De Puytorac *et al.*, 1974).
Sub Class	:	Peritrichia (Calkins 1933).
Order	:	Peritrichida (Stein 1852).
Sub Order	:	Sessilina (Kahl 1933).
Family	:	Epistylididae (Kahl 1933).
Genus	:	Epistylis (Ehrenberg 1832).
Species	:	*niagarae* (Kellicott 1883).

Many lateral protrusions were arised from the prosome between cephalosome and metasome (Figures 18.3 and 18.4). The bossulated abnormal protrusions were more abundant in the nauplii. Bossulated protrusions consisting of bud like structures were observed along the outer periphery of the copepods measured around $1/3^{rd}$ of the size of the entire affected copepods. Copepods with smooth protrusions had a translucent and rounded external surface (Figures 18.5 and 18.6). The histological observations of abnormal protrusions on copepods showed numerous necrotic tissues Figures in 18.7 and 18.8. These necrosis may have been caused by a lesion due to disease or the resistance of blood supply to the tissue of the affected area. Herniated protrusions consisting of necrotic tissue were identified in the affected tissue of copepods (Figures 18.9 and 18.10).

Table 18.1: Appearance of *E. niagarae* on *M. aspericornis* during the sampling period in Velachery lake Chennai.

Months	Prosome of M. aspericornis				Urosome of M. aspericornis		
	Head	Thorax	Antenna	Per cent	Egg Sac	Caudal Style	Per cent
Sep-09	29	21	13	19	31	6	11
Oct-09	23	17	9	15	43	8	16
Nov-09	36	25	12	23	36	16	16
Total	88	63	34	–	110	30	–
Per cent	–	–	–	57	–	–	43

Discussion

The first observation of *E. niagarae* infests on freshwater cyclops *Cyclops vicinus* living in Hazar lake was reported by Saler and Dorucu (2008). We reported here the fortuitous observations of freshwater copepod *Mesocyclops aspericornis* which was heavily infested by peritrichous ciliates namely *E. niagarae*. Our findings could represent that the ciliate protozoans act as epibionts without causing any harmful effect on the host. Saler and Dorucu (2008) also noticed that there was no incidents of infestation occurred on *Cyclops vicinus* by the ciliate protozoan *E. niagarae* from Hazar lake in Turkey.

The absence of *E. niagarae* on nauplii and less percentage of infestation on younger copepodites may be due to the process of moulting where shedding of

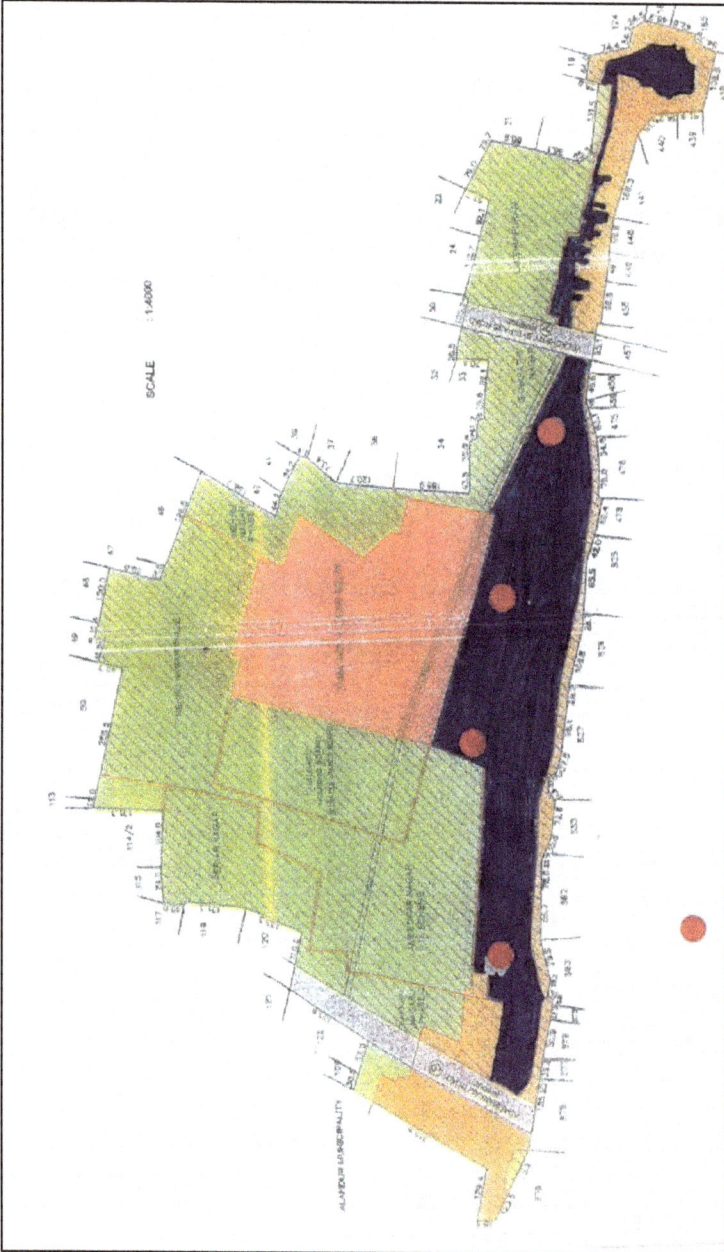

1–4 Sampling locations

Figure 18.1: Map of Lake Velachery showing sampling locations

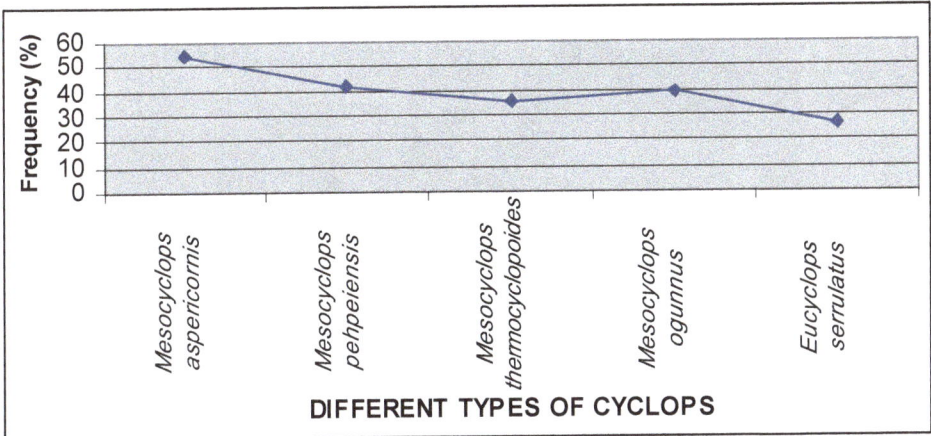

Figure 18.2: Frequency of protrusions in different types of cyclops.

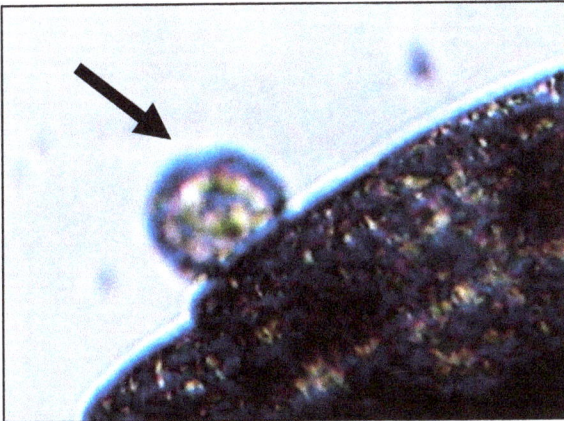

Figure 18.3: *Mesocyclops aspericornis* with an abnormal smooth protrusion with a rounded external surface Line = 0.1 mm.

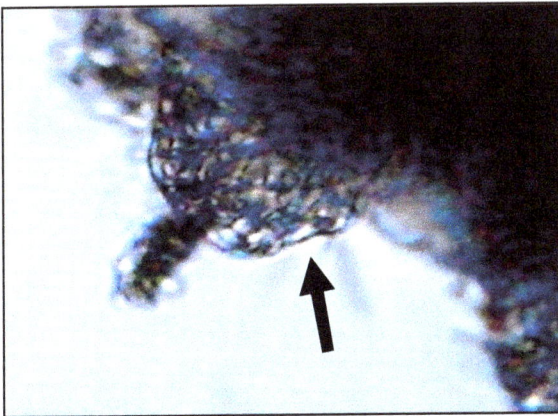

Figure 18.4: *Mesocyclops thermocyclopoides* with a bossulated abnormal protrusion consisting of a bud-like protuberance. Line = 0.1 mm.

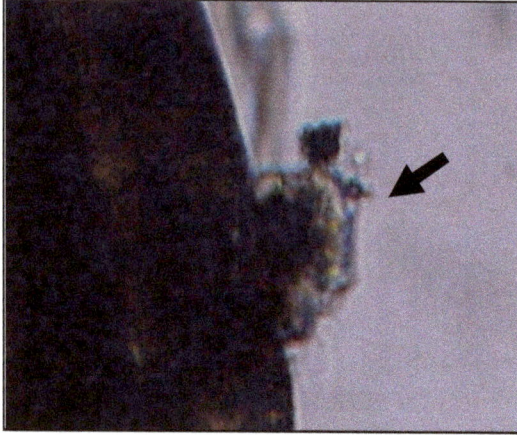

Figure 18.5: *Mesocyclops pehpeiensis* with an amorphous abnormal protrusion on lateral side Line = 0.1 mm.

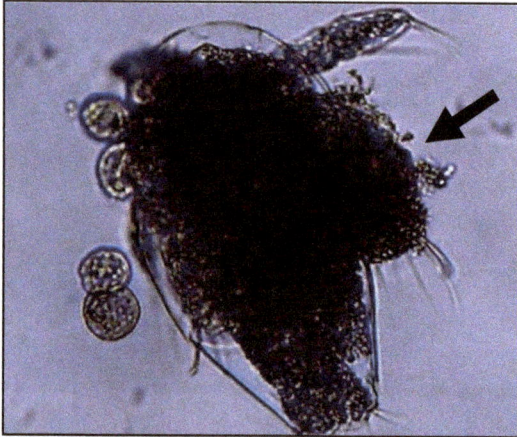

Figure 18.6: Naupliar stage of a copepod with a bossulated abnormal protrusion. Line = 0.1 mm.

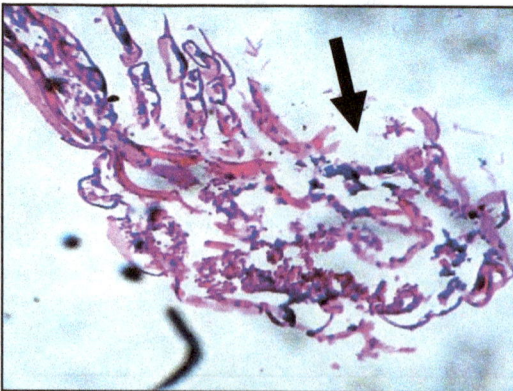

Figure 18.7: Histological section of a copepod with necrotic tissue and contains finely granular material. Line = 100 μm.

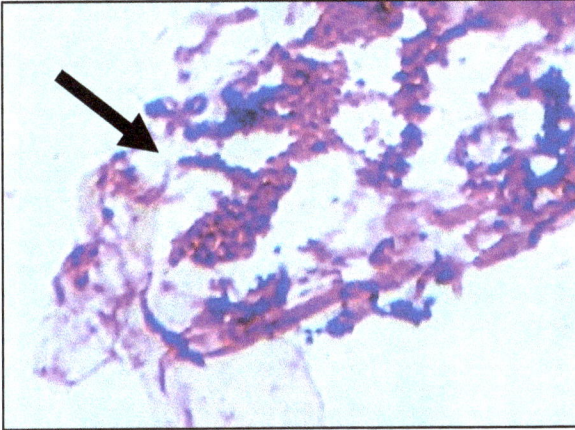

Figure 18.8: Histological section of a copepod showing that the protrusion consists of herniated, non-cellular lipid–like material. Line = 100 µm.

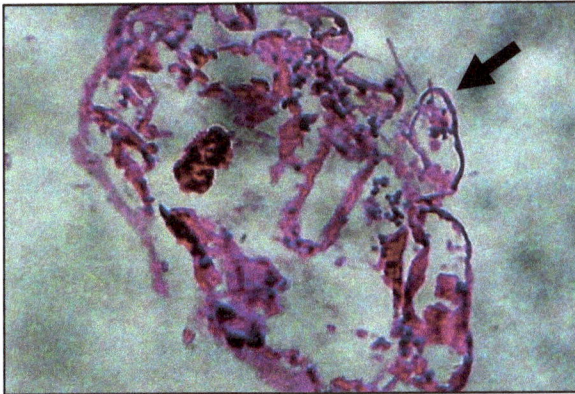

Figure 18.9: Histological section of a copepod showing necrotic tissue. Line = 100 µm.

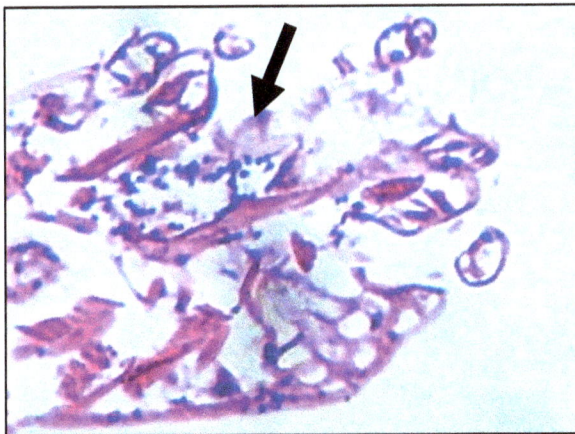

Figure 18.10: Histological section showing of a copepod with a protrusion containing round bodies with no obvious nuclear material. Line = 100 µm.

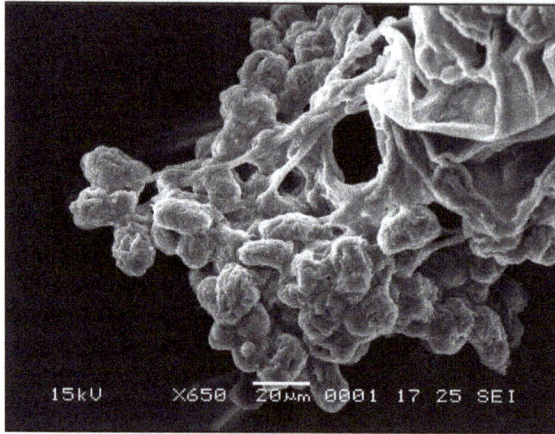

Figure 18.11: View of *E.niagarae* on the head region of *M. aspericornis* as colonial form.

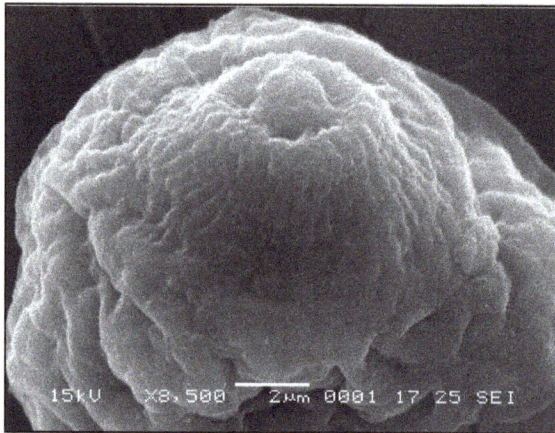

Figure 18.12: *E. niagarae* showing cone shaped body with an enlarged disc shaped peristome with conspicuous cilia on *M. aspericornis*.

Figure 18.13: Epibiont *Epistylis niagarae* observed on the thorax region of *Mesocyclops aspericornis*. Scale bar = 20 μm.

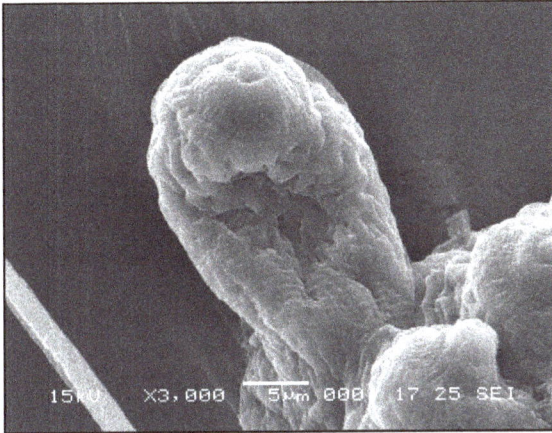

Figure 18.14: Peritrich ciliate *E.niagarae* showing single telotroch (individual).

Figure 18.15: View of *Epistylis niagarae* on the head region of *Mesocyclops aspericornis*. Scale bar = 20 µm.

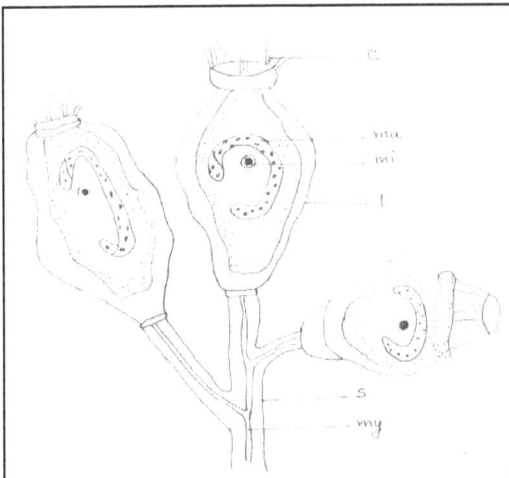

Figure 18.16: Epibiont *Epistylis niagarae*.

carapace removed epibionts during their developmental stages (Overstreet 1983; Gortz 1996). The nauplii of Velachery lake were completely free from epibiont infestation. One of the reasons pointed out by Overstreet (1983) towards infestation on larger animals was the size of the carapace which supports the epibiont to live on. Based on this we concluded that the adult females and males with lengthy body surface area were more susceptible to epibiont infestation than the copepodites and nauplii. The exact reason for epibionts' infestation and its preference needs further study.

Protrusions are composed of necrotic or degenerating tissues (Omair *et al.,* 2001). According to Bridgemen *et al.* (2000); Omair *et al.* (2001) a puncture which was made externally either by the parasites or the predators may initiate the development of abnormal protrusions on copepods. Ellobiopsid parasites were reported to be the first root cause for this abnormal protrusions (Bridgeman *et al.,* 2000). The abnormal protrusions had no obvious nuclear material and were found to be round in shape.

The various types of abnormal protrusions observed in the copepods very clearly revealed the diverse nature of histologic manifestations. No causative agent has been identified so far. Further investigation is essential to determine the etiology for the varied abnormal protrusions seen on the copepod.

References

Bridgeman, T.B., Messick, G.A. and Vanderploeg, H.A., 2000. Sudden appearance of cysts and ellobiopsid parasites on zooplankton in a Michigan lake or potential explanation of tumor-like abnormalities. *Can. J. Fish Aquat. Sci.,* 57: 1539–1544.

Calkins, G.N., 1933. *The Biology of the Protozoa,* 2nd Edn. Lea and Febiger, Philadelphia, pp. 607.

Cheng, C., 1964. *The Biology of Animal Parasites.* W.B. Saunders Company, Philadelphia and London, pp 153–170.

Crisafi, P. and Crescenti, M., 1975. Coseguenze delle attivita umane sullo zooplancton del mare di Taranto. *Boil. Pesca Piscic. Idrobiol.,* 30: 207–218.

Crisafi, P. and Crescenti, M., 1977. Confirmation of a certain correlation between polluted areas and tumor-like conditions as well as tumors growths in pelagic copepods provening from numerous seas of the world. *Rapp. Comm. Int. Mer Medit.,* 24(10): 155.

De Puytorac, P., Batisse, A., Bohatier, J., Corliss, J.O. and Deroux, G., 1974. Proposition d'une classification du phylum Ciliophora Doflein, 1901. *Comptes Rendus de l'Académie de Sciences,* Paris, 278: 2799–2802.

Doflein, F., 1897. Studien Zur Naturgeschichte der Protozoen. I. *Kentrochona nebaliae* Rompel. *Zool. Jahrb. Anat.,* 10: 619–41.

Dragesco, J. and Dragesco-Kerneis, A., 1986. *Cilies libres de l Afrique intertripicale.* Paris, Orstom, pp. 559.

Edmondson, W.T., 1959. *Freshwater Biology*, 2nd Edn. John Wiley and Sons Inc, New York.

Ehrenberg, C.G., 1832. Uber die Entwicklung und Lebensdauer der Infusionsthiere: nebst ferneren Beitragen zu einer Vegleichung ihrer organischen systeme. *Abh. Akad. Wiss. Berlin*, 1831: 1–154.

Fenchel, T., 1987. *Ecology of Protozoa.* Science Tech Publishers, Madison, Wisconsin, pp. 197.

Fernandez-Leborans, G. and Tato-Porto, M.L., 2000. A review of the Species of Protozoan epibionts on Crustaceans. I. Peritrich ciliates. *Crustaceana*, 73: 643–684.

Gortz, H.D., 1996. Symbiosis in Ciliates. In: *Ciliates: Cells as Organisms*, (Eds.) K. Hausmann and P.C. Bradbury. Gustav Fischer, Stuttgart, p. 441–462.

Honigberg, B.M., 1964. A revised classification of the Phylum Protozoa. *Protozool.*, 11: 7–20.

Howard, D.W. and Smith, C.S., 1983. Histological techniques for marine bivalve mollusks. Woods Hole, MA : US Dept. Commerce, NOAA Tech. Memo. NMFS–F/NEC. 25, 97pp.

Kahl, A., 1933. Ciliata Libera et Ecto Commensalia. In: *Die Tierwelt der Nord-und Ostee*, (Eds.) G. Grimpe and E. Wagler. Leipzig.

Kellicott, D.S., 1883. *Epistylis niagarae* sp.n on Crayfish. *Trans. Am. Micr. Soc.*, 5: 110.

Omair, M., Vanderploeg, A., Jude, D. and Fahnenstiel, G., 1999. First observation of tumor-like abnormalities (exophytic lesions) on lake Michigan zooplankton. *Can. J. Fish Aquat. Sci.*, 56: 1711–1715.

Omair, M., Naylor, B., Jude, D., Quddus, J., Beals, T. and Vanderploeg, H., 2001. Histology of herniations through the body wall and cuticle of zooplankton from the Laurentian Great Lakes. *J. Invertebr. Pathol.*, 77: 108–113.

Overstreet, R.M., 1983. Metazoan Symbionts of Crustaceans. In: *The Biology of Crustacea*, (Ed.) D.E. Bliss. Academic Press, London, pp. 155–250.

Saler, S. and Dorucu, M., 2008. The first observation of *Epistylis niagarae* on *Cyclops vicinus* Uljanin, 1875 living in Hazar lake (Elazig-Turkey). *Int. J. of Science and Technology*, 3(2): 161–163.

Silina, N.I. and Khudolei, V.V., 1994. Tumor-like anomalies in planktonic copepods. *Hydrobiol. J.*, 30: 52–55.

Sleigh, M., 1979. *Biologia de los protozoos.* Madrid: H. Blume. pp. 399.

Sprague, V. and Couch, J., 1971. An annotated list of protozoan parasites, hyperparasites and commensals of decapod Crustacea. *J. Protozoology*, 18: 526–537.

Stein, F., 1852. Neue Beitrage Zur Kenntnisder Entwicklungs-geschichte und des feinerer Baues der Infusionsthiere *Z. Wiss. Zool.*, 3: 475–509.

Vanderploeg, H.A., Cavaletto, J.F., Liebig, J.R. and Gardner, W.S., 1998. *Limnocalanus macrurus* (Copepoda : Calanoida) retains a marine arctic lipid and life cycle strategy in Lake Michigan. *J. Plankton Res.*, 20: 1581–1597.

Wahl, M., 1989. Marine epibiosis I Fouling and antifouling some basic aspects. *Mar. Ecol. Prog. Ser.*, 58: 175–189.

2013, Glimpses of Animal Biodiversity

Editor: **Dr. K. Muthuchelian,** *Vice Chancellor, Periyar University, Salem*

Published by: Daya Publishing House, NEW DELHI

Pages **186–193**

Chapter 19

Diversity and Faunastic Studies on Smaller Moths (Insecta: Microlepidoptera) of Western Ghats, Kerala

*R.S.M. Shamsudeen**

Department of Zoology,
Sir Syed College, Kerala

ABSTRACT

The Kerala part of Western Ghats is known to be rich in faunal diversity and no detailed survey of microheteroceran has been made in this region. It was in this context that the present study on this group of insects in the Kerala part of Western Ghats was undertaken to generate baseline data on the fauna of this region and to develop an easy identification scheme. Considering the large number of taxa involved, the scope of the present study has been limited to three super families *viz.,* Tineoidea, Gelechioidea and Yponomeutoidea. Intensive survey of Microheterocera has been made in different parts of Kerala *viz.,* Silent Valley, Mukkali, Muthanga, Sultan Bathery, Amarambalam, Meenmutty, Vellimuttam, Nilambur, Peechi, Vazhani, Sholayar, Thekeddy, Rajamalai, Ranni, Thenmala, Arienkavu, Rosemala, Kattlapara, Achenkovil, Neyyar and Peppara. Sampling of insects was done by light trapping using a lighted sheet. Altogether, 66 species of Microheterocera belonging to the families Psychidae, Tineidae, Oecophoridae, Lecithoceridae, Gelechiidae, Blastobasidae, Cosmopterigidae, Plutellidae, Yponomeutidae, Lyonetiidae, Ethimidae, Glyphipterigidae and Heliodinidae have been recorded in this study. A major share of moths collected

---

* Corresponding Author: E-mail: smartysamsu@gmail.com

in the study belonged to Gelechiidae, Tineidae, Oecophoridae and Cosmopterigidae. The faunal elements were interesting in that they contained several new records for the region–six species for Kerala; 47 for Southern India; two species for India and three species were new to science.

Keywords: *Microlepidoptera, Western Ghats, Tineoidea, Gelechioidea, Yponomeutoidea.*

Introduction

Microheterocera comprises an informal grade of moth taxa. Essentially, it includes a paraphyletic assemblage, of all the primitive as well as higher taxa of the Ditrysian Lepidoptera. Measurement given here are the wingspan of a specimen with the wings spread (or set). Sizes are relative but we describe specimens of 5-12 mm as very small, 9-20 mm, as small, 18-50 mm or more as large (Robinson and Tuck, 1996).

Methodology

The present study was conducted in Kerala part of Western Ghats. The best method of collecting Microlepidoptera was to attract them at night to an illuminated vertical white sheet. The sheet measuring 70 cm × 55 cm touches the ground where it can be nchored with stones. The light source we used was an 18-watt CFL (Compact Fluorescent lamp) powered by a 12-watt car battery. Microheterocera, which rest over the white sheet, were collected in a separate vial, to avoid trampling by other insects. (Shamsudeen *et al.,* 2005).

Results and Discussion

The faunal elements were interesting in that they contained several new records for the region six species for Kerala; 47 for Southern India; two species for India and three species were new to science.

Super-family: Tineoidea

Family: Psychidae

Sub-family: Psychinae

    1. *Brachycyttarus subteralbata* Hampson, Host: Unknown.

       Remarks: Adults were collected from light.

    2. *Eumeta crameri* Westwood

       Host: Unknown. Remarks: Adults were collected from light.

    3. *Metisa plana* Walker

       Host: Serious pest of oil palm.

    4. *Pteroma plagiophleps* Hampson

       Host: *Casuarina equestifolia*

Family: Tineidae

Sub-family: Perissomastictinae

    5. *Edosa glossoptera* Rose and Pathania, 2003

Remarks: New record for South India. Previously it was reported from Himachal Pradesh that was new to Science.

6. *Edosa opsigona* (Meyrick)

Remarks: *Edosa opsigona* (Meyrick) is being reported for the first time from Kerala state in view of its earlier distribution mentioned above (Meyrick, 1911).

Sub-family: Tineinae

7. *Monopis monachella* Hubner

Host: Collected from Tiger scat.

Remarks: In view of the above distribution this is a new record for South India.

8. *Monopis* sp. 1

Host: Collected from Tiger scat.

Remarks: New record for India.

9. *Monopis* sp. 2

Collection data: Wayanad Jul 2002 (2 ex).

Remarks: In view of the above distribution this is a new record for India.

10. *Tinea ixitis* Meyrick, 1894, *Trans. Ent. Soc. London,* 1894: 27.

Remarks: In view of the above distribution this is a new record for South India.

11. *Tinea pellionella* Linnaeus

Remarks: Household pest. Larva feeding on fur, feather, carpets, woolens etc.

12. *Tinea platyntis* Meyrick, 1894

Remarks: In view of the above distribution this is a new record for South India.

13. *Tinea synaema* Meyrick

Remarks: In view of the above distribution *Tinea synaema* is a new record for South India.

Sub-family: Setomorphinae

14. *Setomorpha rutella* Zeller

Host: Larva found on dry tobacco leaves, also from larva on stored coriander seeds, on Wheat flour, and on *Dolichos biflorus*.

Sub-family: Erechthiinae

15. *Erechthias platydelta* Meyrick

Remarks: Adult collected from light trap. New record for South India.

Sub-family: Hieroxestinae

16. *Opogona leucodeta* Meyrick

Remarks: Adult collected from light trap. New record for South India.

17. *Opogona xanthocrita* Meyrick

    Host: Bred from dead wood.

    Remarks: New record for South India.

18. *Opogona* sp.

    Remarks: New record for South India.

19. *Pyloetis mimosae* Meyrick

    Remarks: New record for Kerala.

Super-family: Gelechioidea
Family: Oecophoridae
Sub-family: Xyloryctinae

1. *Nephantis serinopa* Meyrick,

   Remarks: New record to South India.

Sub-family: Statmopodinae

2. *Statmopoda balanarcha* Meyrick

   Remarks: New record for South India. Previously it was reported from Himachal Pradesh that was new to Science.

3. *Stathmopoda theoris* Meyrick

   Host: Larva feeds on flower heads of sun flower (Helianthus), larva reared from cholam ear heads from refuse found in the fork of a tamarind tree, and from palm fibre chewed by *Oryctes rhinoceros;* from ripe fallen gular (*Ficus glomerata*) fruits.

   Remarks: New record for South India.

Sub-family: Oecophorinae

4. *Eucleodora coronis* Meyrick

   Remarks: New record for South India

5. *Periacma plumbea* Meyrick

   Remarks: This species is a new record for South India.

6. *Promolactis semantris* Meyrick

   Host: Bred out from *Shorea robusta* and *Eugenia jambolana.*

   Remarks: New record for South India

7. *Promalactis thiasitis* Meyrick, 1908, *Journ. Bombay. Nat. Hist. Soc.* 18: 807.

   Collection data: Muthanga; October, 2001 (2 ex).

   Distribution: Bihar, Orissa, U.P.

   Host: Sal (*Shorea robusta*), *Eugenia jambolana.*

   Remarks: In view of the above distribution this is a new record for India.

8. *Tonica niveferana* Walker

   Host: Young larva bores on stem of *Bombax malabaricum.*

Sub-family: Hypertrophinae
9. *Eupselia isacta* Meyrick

Remarks: New record for South India.

Family: Ethmidae
Sub-family: Ethmiinae
1. *Ethmia acontias* Meyrick

Host: Bred from *Cynoglossum lanceolatum*.

Remarks: Adult collected from light. New record for South India.

Family: Lecithoceridae
Sub-family: Lecithocerinae
1. *Timyra pastas* Meyrick

Remarks: New record for South India.

2. *Timyra xanthaula* Meyrick

Remarks: Collection was made only from moist deciduous forest (Shendurny and Peechi only). New record for South India.

3. *Lecithocera* sp.

Remarks: Probable for new species.

Sub-family: Torodorinae
3. *Hygroplasta lygaea*

The present survey shows that this species being reported for the first time from South India.

4. *Hygroplasta spoliatella* (Walker), *Gelechia spoliatella* Walker, 1864, *List Specimens lepid. Insects Colln Br. Mus.,* 29: 659.

Remarks: *Hygroplasta spoliatella* (Walker) is being reported for the first time from South India.

5. *Hygroplasta* sp.

Remarks: Only one specimen collected during the study.

Family: Gelechiidae
1. *Anarsia patulella* Meyrick

Remarks: New record for Kerala.

2. *Anarsia isogama* Meyrick

Remarks: New record for Kerala.

3. *Anarsia* sp. 1

Remarks: New record for South India.

4. *Anarsia* sp. 2

Remarks: New record for South India.

5. *Hypatima haligramma* (Meyrick)

Host: *Mangifera indica*.

Remarks: New record for South India.

6. *Sitotroga cerealella* (Olivier)

Host: Larva of *Sitotraga cerealella* feeds on stored grain (rice, maize, etc) and is always a minor and sporadically major pest.

Sub-family: Dichomeridinae

7. *Dichomeris evidantis* Meyrick

Host: The larva rolls the green leaves of *Dalbergia sissu.*

Remarks: This species is a new record for South India.

8. *Dichomeris ianthes* Meyrick

Host: Feeds on Medicago, Cyamopsis and is a pest of indigo.

Remarks: New record to South India.

9. *Dichomeris* sp. 1

Remarks: Probable for new species.

10. *Dichomeris* sp. 2

Remarks: This species is a new record for South India.

Sub-family: Symmocinae

11. *Symmoca signetella*

Remarks: New record for Western Ghats.

12. *Symmoca* sp.

Remarks: New to science.

Sub-family: Anacampsinae

13. *Fresilia* sp.

Remarks: Probable for new species.

14. *Idiophantis acanthopa* Meyrick

Remarks: New record for South India.

15. *Onebala hibisci* Stainton

Remarks: New report for South India.

16. *Onebala hoplophora* Meyrick

Remarks: This species is a new record for South India.

17. *Stegasta* sp.

Remarks: New record for Kerala.

Family: Blastobasidae
Sub-family: Blastobasinae

1. *Blastobasis pulverea* Meyrick

Remarks: Species is new record to South India.

2. *Blastobasis* sp.

Host: Unknown.

Remarks: New record for South India.

    3. *Cladobrostis* sp. 2

    Remarks: New report for the state.

Family: Cosmopterigidae
Sub-family: Cosmopteriginae

    1. *Cosmopterix mimetis* Meyrick

    Host: Host of Casuarina.

    Remarks: First time from South India.

    2. *Labdia semicoccinea* Stainton, Life histories of Indian Insects, Microlepidoptera. *Mem. Dep. Agric., India.* Vol. VI, No. 1: 100.

    Host: Reared from stems of *Cajanus indicus.*

    Remarks: This species is a new record for Kerala.

    3. *Labdia stibogramma* Meyrick, *A field guide to smaller moths of South East Asia. Malaysian Nature Society, Malaysia.* 1994:70.

    Host: Larva is generalized feeders on plant detritus. This species is a new record for India.

    4. *Labdia xylinaula* Meyrick

    Remarks: New record for South India.

    5. *Limnaecia chromaturga* Meyrick

    Remarks: New record for South India.

    6. *Limnaecia peronodes* Meyrick

    Remarks: New record for South India.

Sub-family: Chrysopeleiinae

    7. *Eumenodora tetrachorda* Meyrick

    Remarks: New record for South India.

    8. *Stagmatophora faceta* Meyrick

    Remarks: New record for South India.

Super-family: Yponomeutoidea
Family: Glyphipterigidae

    1. *Phycodes minor* Moore

    Host: Larva has been reared from *Ficus carica, F. heterophylla.*

Family: Plutellidae
Sub-family: Plutellinae

    1. *Plutella xylostella* Linnaeus

    Host: Pest of cabbage, Cauliflower, radish, mustard and other cruciferous plants.

Family: Yponomeutidae

    1. *Argyresthia* sp.

    Collection data: Vazhani and Sholayar, July 2003.

Remarks: New record to South India. Adult collected from light in MDF habitat.

Family: Attevidae

Sub-family: Attevinae

1. *Atteva fabriciella* Swederus

   Host: *Ailanthus exclesa.*

   Family: Lyonetidae

Sub-family: Cemiostominae

1. *Leucoptera sphenograpta* Meyrick

   Host: Larva mining blotches in the leaves of *Dalbergia sisso.*

   Remarks: New report for South India.

Family: Heliodinidae

1. *Eretmocera* sp.

   Remarks: New record for South India.

Acknowledgements

MoEF, KFRI, Peechi, Gaden Robinson, BNHM, K.T. Park, Kangwon University, Korea.

References

Robinson, G.S. and Tuck, K.R., 1996. A revisionary checklist of the Tineidae (Lepidoptera) of the Oriental region. *Occ. Pap. Syst. Ent.,* 9: 29.

Shamsudeen, R.S.M., Chandran, R., and Mathew, G., 2005. Collection of Microheterocera: A Newer Method. *Bug 'R' All,* 8(1): 3.

2013, Glimpses of Animal Biodiversity   *Pages* **194–198**

Editor: Dr. K. Muthuchelian, *Vice Chancellor, Periyar University, Salem*
Published by: Daya Publishing House, NEW DELHI

Chapter 20

Diversity and Epidemiological Surveilance of Dengue in *Aedes* Mosquitoes: A Report from Virudhunagar District

R. Sindhupriya[1], M. Paulpandi[2] and S. Ilango[1]
[1]Post Graduate and Research Department of Zoology,
Ayya Nadar Janaki Ammal College (Autonomous), Sivakasi – 626 124
[2]Department of Genomic Research, Barathiar University, Coimbatore

ABSTRACT

Dengue fever (DF), dengue hemorrhagic fever (DHF) and dengue shock syndrome (DSS) are re-emerging infectious diseases caused by the four serotypes of dengue virus (DENV 1 to 4) belonging to the family *Flaviviridae* and genus *Flavivirus*. In the absence of a safe, effective prevention and control of dengue outbreaks depend upon the surveillance of cases and mosquito vector. An outbreak of dengue virus in *Aedes* was investigated during first week of June 2008 to December 2009; suspected cases of DF were reported from many places in rural and urban areas of Virudhunagar district. A total of 676 female adult *Aedes* mosquitoes were collected from the field in order to serve as an early warning monitoring tool for dengue outbreaks. DENV infected and uninfected *Aedes* were screened by dengue IgM capture assay and Heamoagglutination assay. A total of 231 mosquitoes were DEN positive by either dengue IgM capture assay (n = 155) or Heamoagglutination assay (n = 112) or both (n = 32). High rainfall and humidity with the temperature range from 22°C to 36°C during the month of June to December might have favored to the breeding and transmission of *Aedes,* in consequence leading to an increase in the number of dengue mosquitoes.

Keywords: *Dengue detection, Transmission and Dengue IgM capture assay.*

Introduction

Dengue virus, which belong as a (+)-stranded encapsulated RNA virus to the family *Flaviviridae,* genus *Flavivirus,* is divided in the four serotypes (DEN-1 to DEN-4). The most prominent vector *Aedes aegypti* and *Aedes albopictus* are widespread in tropical and subtropical countries especially in urban areas. The global prevalence of dengue has risen dramatically and dengue is now endemic in more than 100 countries. DENV is caused dengue hemorrhagic fever (DHF) by hemorrhagic signs in infected cells and is also caused dengue shock syndrome (DSS) and is characterized by features of DHF plus evidence of circulatory failure manifested by hypotension or hypertension, cold clammy skin and restlessness (Fauci *et al.,* 2005; Butt *et al.,* 2008). The four serotypes of virus are transmitted to humans through the bites of infected female mosquitoes (*Ae. aegypti* and *Ae. Albopticus).* DENV prevention and control of the disease mainly depends on the epidemiological surveillance of cases and mosquito vectors. The vector surveillance allows timely implementation of emergency mosquito control measures to kill adults and destruction of breeding places to limit an impending outbreak. Such surveillance based on the isolation and identification of DENV in *Aedes* female infecting the human population provides an important means of early detection for dengue outbreaks (Thenmozhi *et al.,* 2007). Therefore, the present study was undertaken to detect the DENV (flavivirus) isolation from field caught female *Aedes* mosquitoes at Virudhunagar district and risk areas are identified using immunological techniques.

Materials and Methods

Study Area

The study area (1652 sq. km) lies between 8° 27' 10" N–18° 20' 82" N latitudes and 74° 16' 07" E–80° 11' 54"E longitudes. This area receives rainfall, *i.e.,* Southwest (June–September) and Northeast (October–December) monsoon. The rainfall is moderate to high, with an average of 800 mm to 850 mm annually. The mean annual temperature is 26°C–28.5°C. The literacy rate is 80 per cent. A network of government-run Primary Health Centres (PHCs) in rural areas and General Hospitals in urban areas serves the health needs of the people. Apart from dengue fever caused by DENV and transmitted by *Aedes,* vector-borne diseases prevalent in the region are dengue and malaria etc.

Sampling Design

Based on the occurrence of mosquitoes breeding sites, the areas are identified as dengue infection risk area and non risk area. The availability of breeding sites is to favor the increase the population of dengue vector *Ae. aegypti* as well as *Ae. albopictus.* Therefore, the selected site was divided into 15 x 15 km grids, so that each grid would contain at least one villate (rural) or ward (urban). Since a total 25 and 10 sites (intersection points–villages/wards) were selected on a systematic random basis, to minimize the cost of survey but at the same time ensured a fair spatial representation of the area for risk and non-risk zones respectively. In the non-risk zone, all the villages/wards falling on either intersection points or nearer to them were selected for the survey, since the negative estimation in the site was more critical from the application point of view. A totally 35 sites were surveyed from the study area on a systematic random basis.

Samples Collection

Adult female *Ae. aegypti* and *Ae. albospictus* were collected from inside houses with the help of mouth aspirators, aided by torch lights. Most of the human dwellings were permanent houses. Outdoor resting habitats consisted of field grasses, cotton fields etc. All adult mosquitoes were anesthetized and identified to species by using microscopic examination and relevant taxonomic references.

Samples Screening

The collected female *Ae. aegypti* and *Ae. albopictus* were dissected individually to separate head, thorax and abdomen. The thorax was triturated with 1 ml mosquito tissue grinders in a solution of 20 per cent acetone extracted normal human serum in PBS, pH 7.4. The suspension was directly screened against DENV, using the ELISA techniques and Heamoagglutination assay. The procedure for antigen capture ELISA was carried out as previously described, with modification (Che *et al.*, 2004). In brief, micro well plates were coated with 100 µ1/well of polyclonal antibodies overnight at 4°C, and then the wells were incubated with a blocking reagent. After removal of the blocking solution, a series of diluted 100 µ1/well of TMB solution was added, and the reaction was stopped after incubation for 10 min with 100 µ1/well of diluted HRP-conjugated MAb was added washing, 100 µ1/well of TMb solution was added, and the reaction was stopped after incubation for 10 min with 100 µ1/well of 1 N sulfuric acid and absorbance was read at 450 nm in a microplate reader.

Results and Discussion

In this study, the adult female mosquitoes were subjected for DENV detection by ELISA. Specimens were considered positive if their optimum density values were greater than negative control values. A total of 676 samples were screened for DENV infection by ELISA from the selected 35 sites. These samples were tested as described above and the results are given in Table 20.1. A 103 samples tested from "non-risk" area were negative for DENV and 573 samples were tested from risk area. A total of 231 mosquito samples were positive for DENV recorded from the "risk" and non-risk areas in Tables 20.1 and 20.2. The present results were also confirmed that showed positive to the DENV by immunological assay (Ilango and Paulpandi, 2009).

Table 20.1: The number of field caught female *Aedes* positive for DENV from selected sites by ELISA.

Risk Status	No. of Sites Selected	No. of Sites Positive for DV	No. of Individuals Tested		
			Positive	Negative	Total
"Non-risk"	20	1	4	99	103
"Risk"	15	12	151	422	573
Total	35	13	155	521	676

Table 20.2: The number of field caught female *Aedes* positive for DENV selected sites by heamoagglutination assay.

Risk Status	No. of Sites Selected	No. of Sites Positive for DV	No. of Individuals Tested		
			Positive	Negative	Total
"Non-risk"	20	1	2	101	103
"Risk"	15	12	110	463	573
Total	35	13	112	564	676

Transmission of DENV is mainly influenced by demographic factors *viz.*, population density, occupation, literacy, migration, economic status and health seeking behavior on the occurrence of mosquito-borne diseases at micro level has been described (Sherchand *et al.,* 2003; Galvez, 2003; Sabesan *et al.,* 2006). Similarly, the vector abundance may be depending on the geo-physical and human associated factors, but the vector survival and the virus transmission are greatly determined by the environmental factors at macro level. In the present study, agree with the each of the geo-environmental risk factors was not taken as a separate entity, but all factors have been considered together a composite index (FTRI), since a combination of factors are responsible for transmission of DENV (Brooker and Michael, 2000).

In the present results was indicated that the area with "non-risk" of DENV transmission could be identified at macro level using immunological assay, particularly developing countries like India. A total area of Virudhunagar region is not fully surveyed even one time and their risk status not known. However, through this attempt, it can be possible to identify the positive cases from the potential "risk" area with any of the existing tools. Therefore, the immunological assay is an alternative tool for epidemiological surveillance for DENV in mosquitoes. Finally, dengue is a man-made problem only and that by the above changes these will be source reduction in mosquito breeding and complete elimination of DENV from human circulation. Therefore, prevention and control of DENV is beset with many challenges and requires high political commitment, multisectoral collaboration and a coordinated community effort to increase awareness about how to control the mosquito that transmits it. Residents should be made responsible for keeping houses and surroundings free from mosquito breeding by emptying and scrub drying the rotate containers once a week.

Acknowledgements

The present work was supported by the management, ANJA College, Sivakasi. The author would like to thank Dr. J. Muthukrishnan, Professor of Environmental Biology, Madurai Kamaraj University for the methodological and statistical processing of the results.

References

Brooker, S. and Michael, E., 2000. The potential of geographical information systems and remote sensing in the epidemiology and control of human helminth infections. *Advances in Parasitol.,* 47: 246–288.

Butt, N., Abbassi, A., Munir, S.M., Ahmad, S.M. and Sheikh, Q.H., 2008. Haematological and biochemical indicators for the early diagnosis of dengue viral infection. *J. Physicians and Surgeons,* 18: 282–285.

Che, X.Y., Qiu, L.W., Pan, Y.X., Wen, K., Hao, W., Zhang, L.Y., Wang, Y.D., Liao, Z.Y., Hua, X., Cheng, V.C. and Yuen, K.Y., 2004. Sensitive and specific monoclonal antibody-based capture enzyme immunoassay for detection of nucleocapsid antigen in sera from patients with severe acute respiratory syndrome. *J. Clin. Microbiol.*, 42: 2629–2635.

Fauci, A.S., Touchette, N.A. and Folkers, G.K., 2005. Emerging infectious diseases: A 10–year perspective from the National Institute of allergy and Infectious Diseases. *Emerg. Infect. Dis.,* 11: 519–525.

Galvez, T.Z., 2003. The elimination of Lymphatic filariasis: A strategy for poverty alleviation and sustainable development: Perspectives from the Phillippines. *Filaria Journal,* 2: 5.

Ilango, S. and Paulpandi, M., 2009. Serological detection of dengue virus in urban areas of Sivakasi Taluk, Tamil Nadu. *Proceedings of Conference on Recent Advances in Zoology,* pp. 10–13.

Sabesan, S., Raju, H.K., Srividya, A. and Das, P.K., 2006. Delimitation of lymphatic filariasis transmission risk areas: A geo-environmental approach. *Filaria Journal,* 5: 12.

Sherchand, J.B., Obsomer, V., Das, T.G. and Hommel, M., 2003. Mapping of lymphatic filariasis in Nepal. *Filaria Journal,* 2: 9.

Thenmozhi, V., Hiriyan, J.G., Tewari, S.C., Samuel, P.P., Paramasivan, R., Rajendran, R., Mani, T.R. and Tyagi, P.K., 2007. Natural vertical transmission of dengue virus in *Aedes albopictus* (Diptera: Culicidae) in Kerala, a Southern India State. *Jpn. J. Infect. Dis.,* 60: 245–249.

2013, Glimpses of Animal Biodiversity
Editor: **Dr. K. Muthuchelian,** *Vice Chancellor, Periyar University, Salem*
Published by: **Daya Publishing House, NEW DELHI**

Pages **199–206**

Chapter 21

Diversity and Foraging Behaviour of Birds in Southern Tropical Thorn Forests of North Eastern Slopes of the Nilgiri Foothills

Vijayakrishnan Sreedhar, A. Selvaraju, A. Balasubramanian, and Kamdi Hemant Bhaskar*

*Forest College and Research Institute,
Tamil Nadu Agricultural University, Mettupalayam – 641 301*

ABSTRACT

Bird diversity study was taken up in the tropical thorn forests of North Eastern slope of Nilgiri Foothills, to assess the different migratory and residential bird populations in the forest. The study result elucidated 151 different bird species at the end of 2 years of continuous field study. Among the total of 151 species of birds recorded 63 species are resident and 88 species are migratory. The study was taken up deep inside the forest using GPS transect survey on systematic random sampling method. Among the migratory birds, the most common visitors recorded were Common Grey Hornbill, Asian Paradise Flycatcher, Pompadour Green Pigeon, Emerald Dove, Changeable Hawk Eagle etc. Long distance migratory birds like Changeable Hawk Eagle was found to visit the study site for breeding and nesting activities during March to April. Residential birds like Indian pitta, Honey Buzzard, Blue bearded bee eater, Red

* Corresponding Author: E-mail: elephanttracker@yahoo.co.in turtle dove etc are very unique to this ecosystem. During the study, attempt was made to analyse the forage behaviour of resident and migratory birds. Among the 151 species of birds, 32 numbers of birds were found to be frugivorous and 119 species omnivorous. The study assessed the seasonal variation of both migratory and residential birds. The population counts of birds were subjected for different diversity indices. The study also observed a unique and specific relationship between birds and trees in the forest. Especially, migratory species such as the common grey hornbill was found to be dependent on *Ailanthus excelsa* tree for nesting and foraging activities. Similarly Pompadour green pigeon, Plum headed parakeets etc, were found to be associated with *Ficus bengalensis*. The population of Pompadour green pigeons and Plum headed parakeets were found to be very high during the fruiting season of *F.bengalensis* in the month of July to August. A better understanding of association between avian fauna and endemic flora will help in studying the importance of birds in refurbishment of the Western Ghats vegetation, one of the world's richest biodiversity hotspots.

Introduction

The Avifaunal diversity of tropical thorn forests of Nilgiri foothills has remained unexplored for time immemorial. Hence, a study was taken up along the North Eastern slopes of Nilgiri foothills, to assess the diversity of migratory and resident bird populations in the forest. A total of 151 species of birds were recorded during two years of field study. The role of birds in restoring degraded landscape is crucial as they disperse seeds of various invasive and native plants after feeding on the fruit, thereby aiding natural regeneration of the flora. Hence the study also focused on the foraging aspects of birds in the study area. Based on their feeding patterns, birds may be classified as frugivorous (fruit eating), carnivorous (flesh eating), piscivorous (fish eating), insectivorous (insect eating) and omnivorous (both plants and animals).

Distinctive symbiotic relationships normally exist between avifauna and flora. Further studies may be plotted for discernment of these relationships as it will definitely assist in studying the importance of birds in restitution of vegetation, especially in degraded landscapes.

Materials and Methods

Study Area

The study was carried out in the natural reserve forests (2 km², 11°30' N and 76°56' E), along the North Eastern slopes of the Nilgiri foothills, in the Tamil Nadu state of Southern India. The altitudes of study area varied from 320 m to 350 m above Mean sea level. The area enjoys both South west and North east monsoon, together contributing a total rainfall of 915 mm per annum. The vegetation is typical thorn comprised of *Gyrocarpus, Dichrostachys cinerea, Erythroxylon monogynum, Randia dumetorum, Hugomia mystax, Ziziphus* sp., *Carissa spinarum,* etc. This forest acts as an excellent habitat for vivid avian, mammalian and herpeto fauna and also acts as a major corridor for elephants along the Nilgiri foothills. The present study emphasized on diversity of avian fauna alone.

Table 21.1: Geographical position and length of transects surveyed.

Transect No.	Starting GPS Point	Ending GPS Point	Length of Transect (km)
1.	11°19.57'N 76°56.20'E	11°19.95'N 76°56.18'E	0.64
2.	11°20.01'N 76°56.24'E	11°20.19'N 76°56.46'E	0.8
3.	11°20.18'N 76°56.48'E	11°20.03'N 76°56.49'E	0.32
4.	11°20.03'N 76°56.52'E	11°20.05'N 76°56.77'E	0.48
5.	11°19.79'N 76°56.75'E	11°19.73'N 76°56.44'E	0.5
6.	11°19.54'N 76°56.45'E	11°19.56'N 76°56.20'E	0.5

Methodology

Population Count

The method employed in surveying the bird diversity was GPS transect survey on a systematic random sampling basis. Six permanent transect lines of 0.3–0.8 km each were laid in different zones within the study site and walked on a regular weekly basis during 2008 and 2009.

The population estimations of the birds were carried out during early hours of the day (06.00 to 07.30) and in the evenings (17.00h to 18.30h). The survey was carried out along all six transects simultaneously to minimize any bias arising from the variations in bird activities.

Foraging

To analyse the foraging behaviour of birds, mostly frugivorous, fruiting trees were located within the study area and the foraging behaviour in terms of bird-tree species and vice versa relationship was studied. Separate data sheets were prepared for taking the population count of birds along the transect lines and at their foraging sites (*i.e.*, Fruiting trees). A correlation study was made between frugivorous population vs. fruiting season.

Diversity Analysis

Population Study

The density and frequency of major resident birds were calculated

Density

It is expressed as number of organisms per unit area and is given by the following formula:

$$D = \frac{\text{Total number of individuals of a species in all sampling units}}{\text{Total number of sampling units sampled}}$$

where,

    D: Density

Frequency

    It is defined as the chance of an individual of a given species to be present in a randomly placed quadrat or transect. It is given by the formula,

$$F = \frac{\text{No. of sampling units in which a species occurs}}{\text{Total number of sampling units sampled}} \times 100$$

where,

    F: Frequency

Diversity Analyses

    Shannon-Weiner and Simpson's index for dominant resident birds were calculated using the number of these birds observed in all 6 transects.

    Shannon-Weiner index, $H = \Sigma\ P_i \ln P_i$

    Simpson's index, $\lambda = \Sigma\ P_i^2$

where, P_i is the population of ith species.

Results and Discussion

    From the data obtained during two years of intense field study, diversity indices (Shannon-Weiner and Simpson's indices), density and frequency values for dominant bird species of the study area were calculated which are as shown in the table. Among the 63 residential bird species, 10 most dominant species were only subjected for diversity analysis. The study result elucidated that Spotted dove (*Streptopelia chinensis*) showed a very high density of 13.5 followed by Jungle crow (*Corvus macrorhynchos*), with a density value of 10.5 and Small Sun bird (*Nectarinia minima*) having a density value of 9.83. Asian koel (*Eudynamys scolopacea*) registered a very low density of 2.17 followed by Indian Pea fowl (*Pavo cristatus*), density value 2.83. The above results are described in Table 21.1 and Figure 21.2. Most of the birds considered for the diversity analysis were found to be present in all 6 transects, thereby showing a frequency of 100 per cent except for Small green bee eater (*Merops orientalis*) and Asian koel (*Eudynamys scolopacea*), present only in 5 out of 6 transect, and having frequency values of 83.33 per cent each.

    The diversity analysis of the birds revealed a very high Shannon-Weiner Index and Simpson's Index of 219.58 and 1313 respectively were recorded in Spotted dove (*Streptopelia chinensis*). On the contrary, very low Shannon-Weiner and Simpson's index values were observed in Asian koel (*Eudynamys scolopacea*).

    Observation of foraging behaviour of birds, primarily frugivorous birds, elucidated some unique relationship between certain species of birds and specific tree species in the forest. For instance, the common grey hornbill (*Tockus birostris*) was found to be associated with *Ailanthus excelsa* tree for its foraging and nesting activities. Similarly the Changeable Hawk Eagle, a slender forest eagle species distributed almost

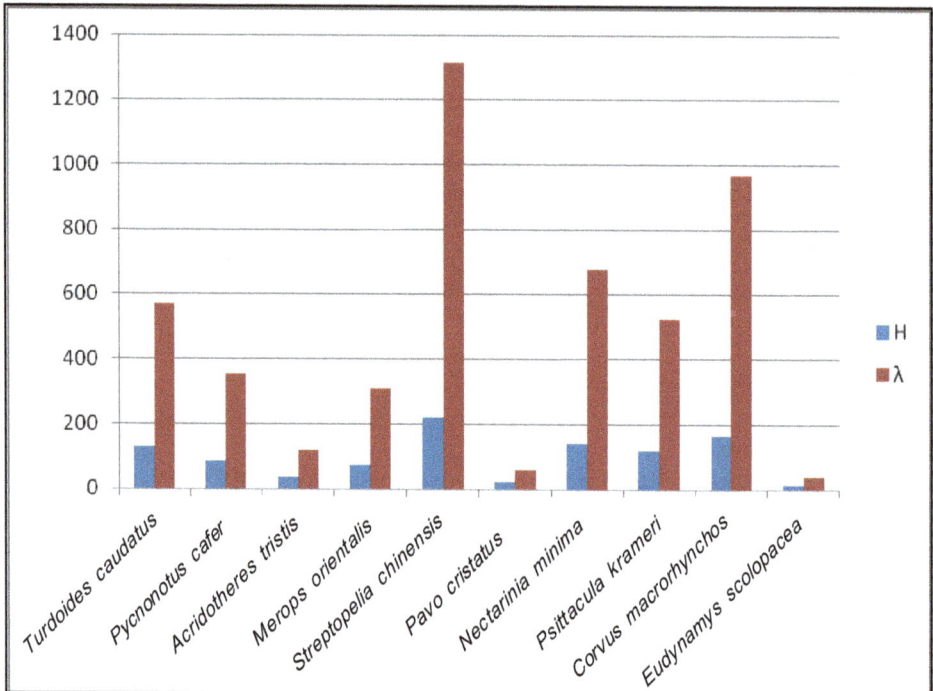

Figure 21.1: Shows Shannon-Weiner and Simpson's indices' values of ten common species observed.

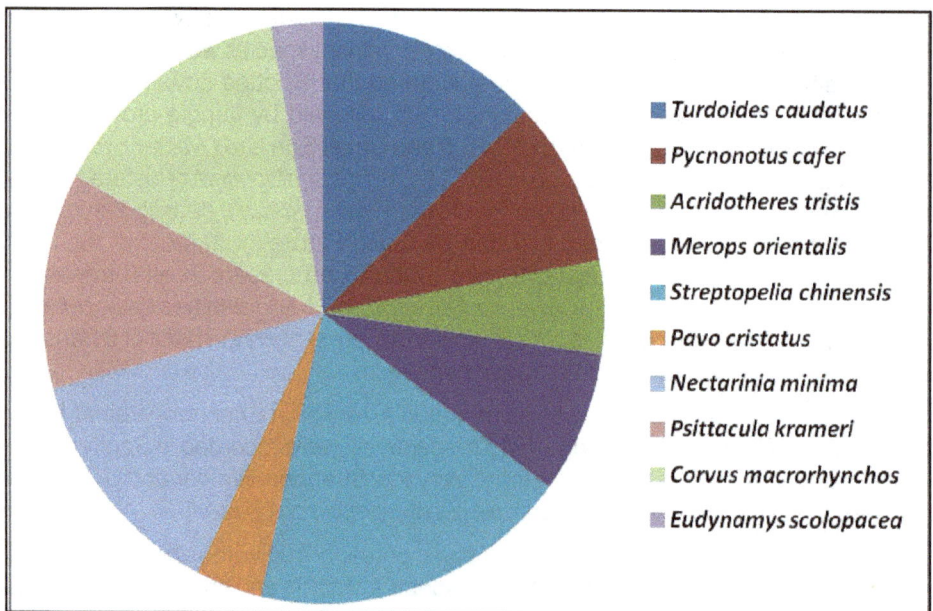

Figure 21.2: D (Density)

throughout the Indian Sub-continent, was found to be consociated with White babul (*Acacia leucophloea*) tree for nesting during the months of March-April.

Table 21.2: Showing the Density, Frequency, Shannon-Weiner and Simpson's index values for population count of dominant bird species in the study area.

Species	Density (D)	Frequency (F)	Shannon-Weiner Index (H)	Simpson's Index (λ)
Turdoides caudatus	9.33	100	127.55	570
Pycnonotus cafer	6.83	100	84.46	353
Acridotheres tristis	3.83	100	34.91	119
Merops orientalis	5.83	83.33	72.65	309
Streptopelia chinensis	13.50	100	219.58	1313
Pavo cristatus	2.83	100	19.66	59
Nectarinia minima	9.83	100	139.88	675
Psittacula krameri	8.83	100	118.53	523
Corvus macrorhynchos	10.50	100	162.45	967
Eudynamys scolopacea	2.17	83.33	13.52	39

Correlation study attempted during the survey showed that the populations of migratory birds were totally dependent on fruiting season of specific tree species. To substantiate this, population count of three frugivorous migratory bird species were

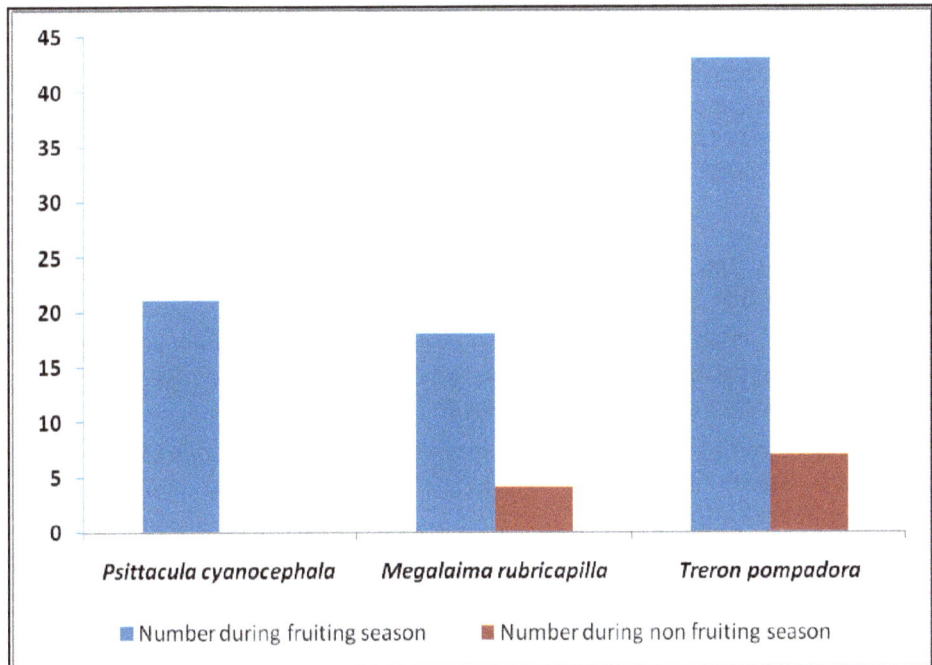

Figure 21.3

recorded during fruiting and non fruiting season of Fig (*Ficus bengalensis*). The population counts were comparatively very low during non fruiting season, which threw light on the fact that there exists a perfect correlation between the population of these frugivorous birds and the fruiting season of trees (Figure 21.3).

Study was done on foraging behaviour of birds to analyse the approximate number of resident and migratory frugivorous species. Out of the total number of 151 species, 32 were found to be frugivorous and the rest 119 carnivorous, piscivorous, insectivorous or omnivorous. Frugivorous birds were given prominence as the role played by them in seed dispersal and initiating natural regeneration of tropical and sub tropical tree species is inevitable. They have been recognized as the main dispersal agent of many invasive plant species (Glyphis *et al.*, 1981; Buchanan, 1989; Dean and Milton, 2000; Stansbury, 2001; Renne *et al.*, 2002).

Aspects of fruit morphology, such as fruit and seed size, seed load (ratio of indigestible seeds to pulp) and seed geometry, affect choices by birds (Herrera, 1984; Howe, 1989; Murray *et al.*, 1993). Smaller fruits predominate among both indigenous and invasive assemblages of vertebrate-dispersed plants and are accessible to a larger variety of fruit eating birds (Green, 1993; Kitamura *et al.*, 2002). Plants with large fruit have comparatively few dispersers in both native and invasive bird-dispersed plants. A related observation from the study area was the increase in natural regeneration of Sandal tree (*Santalum album*) which may be attributed to Asian koel (*Eudynamys scolopacea*) as they are major predator of sandal fruits. Seeds voided by birds have increased chances of regeneration as they undergo scarification by gut acids (Traveset *et al.*, 2001).

This study attempted to analyse the diversity of avifauna along North Eastern Slopes of Nilgiri foothills. Also, it analysed the foraging patterns of selected frugivorous bird species. A more detailed study may be plotted to judge the symbiotic relationship between various bird species and native and invasive flora as it will contribute towards role of birds in rejuvenating landscapes that are in the verge of degradation.

References

Ali, Salim, 1979. *The Book of Indian Birds*. Oxford University Press, Walton Street, Oxford, pp. 354.

Buchanan, R.A., 1989. Pied currawongs (*Strepera graculina*): Their diet and role in weed dispersal in suburban Sydney, New South Wales. In: *Proceedings of the Linnean Society of New South Wales*, 111: 241–255.

Dean, W.R.J. and Milton, S.J., 2000. Directed dispersal of *Opuntia* species in the Karoo, South Africa: are crows the responsible agents? *Journal of Arid Environments*, 15: 305–311.

Glyphis, J.P., Milton, S.J. and Siegfried, W.R., 1981. Dispersal of *Acacia cyclops* by birds. *Oecologia*, 48: 138–141.

Gosper, C.R., Stansbury, C.D. and Vivian-Smith, G., 2005. Seed dispersal of fleshy-fruited invasive plants by birds: Contributing factors and management options. *Divers. Distrib.*, 11: 549–558

Green, R.J., 1993. Avian seed dispersal in and near subtropical rainforests. *Wildlife Research*, 20: 535–557.

Herrera, C.M., 1984. A study of avian frugivores, bird-dispersed plants, and their interaction in Mediterranean scrublands. *Ecological Monographs*, 54: 1–23.

Howe, H.F., 1989. Scatter and clump-dispersal and seedling demography hypothesis and implications. *Oecologia*, 79: 417–426.

Kitamura, S., Yumoto, T., Poonswad, P., Chuailua, P., Plongmai, K., Maruhashi, T. and Noma, N., 2002. Interactions between fleshy fruits and frugivores in a tropical seasonal forest in Thailand. *Oecologia*, 133: 559–572.

Murray, K.G., Winnett-Murray, K., Cromie, E.A., Minor, M. and Meyers, E., 1993. The influence of seed packaging and fruit color on feeding preferences of American robins. *Vegetatio*, 107/108: 217–226.

Renne, I.J., Barrow, W.C., Randall, L.A.J. and Bridges, W.C., 2002. Generalized avian dispersal syndrome contributes to Chinese tallow tree (*Sapium sebiferum*, Euphorbiaceae) invasiveness. *Diversity and Distributions*, 8: 285–295.

Stansbury, C.D., 2001. Dispersal of the environmental weed bridal creeper *Asparagus asparagoides* by silvereyes *Zosterops lateralis* in south-western Australia. *Emu*, 101: 39–45.

Traveset, A., Riera, N. and Mas, R.E., 2001. Passage through birdguts causes interspecific differences in seed germination characteristics. *Functional Ecology*, 15: 669–675.

2013, Glimpses of Animal Biodiversity
Editor: Dr. K. Muthuchelian, *Vice Chancellor, Periyar University, Salem*
Published by: Daya Publishing House, NEW DELHI

Pages **207–211**

Chapter 22

Status of Testes in Relation to Total Length (TL) and Gonadosomatic Index (GSI) in Freshwater Fish *Mystus cavasius* (Ham.) from Bhadra Reservoir, Western Ghats, Karnataka

H.M. Ashashree, J.D. Kashinath, M. Venkateshwarlu and H.M. Renuka Swamy*

*Sahyadri Science College (Autonomous),
Shimoga Dist., Karnataka*

ABSTRACT

In the present investigation study of testes in relation to Gonadosomatic Index (GSI) and Total Length (TL) of freshwater fish, *Mystus cavasius*, have been studied from Bhadra region. The minimum GSI were observed in the month of December, while the maximum GSI were recorded in the month of July that may be attributed towards the peak period of spawning. The GSI provides an idea regarding the different spawning seasons of the present fish from the study

---

* Corresponding Author: E-mail: asha.chinku.shree@gmail.com

area. Similarly the maximum GSI of the fishes were observed in the higher length group of fishes *i.e.* 210-220 mm and minimum GSI were recorded in the lower length group *i.e.* 140-150 mm. In the present study, GSI is found to be directly proportional to spawning season and inversely proportional to post-spawning season and it does not show any correlation with total length in the present fish.

Keywords: *Mystus cavasius, Gonado somatic index, Spawning season, Total length.*

Introduction

The Bhadra reservoir is situated at Lakkavalli village, Tarikere taluk, Chikkamagalore district of Karnataka. It is situated at an elevation of 601 mts above MSL. The dam is located at latitude 13° 42' 00" N and longitude 75° 38' 20" E. The Bhadra basin gets the rainfall from 117 cm to 513 cm and the temperature varies from 30.14 to 18.76°C. The reservoir is having 56.89 mts depth. This is a multipurpose project.

Knowledge on biology of fish breeding is extremely important for the development of fish culture. Maturity and spawning are important biological aspects to be studied in fishes as it is useful in its management (Anju and Anoop, 2003). Reproductive cycle and gametogenesis are important parameters in understanding the reproduction of native fish species and to the establishment of conservation programmes (Bazzoli and Godinho, 1991). *Mystus cavasius* is a commercially important fish having high protein content and good market value. This also forms an important fishery along Bhadra reservoir. The culture practice of above fish is not reported from the area this is due to the lacking of the proper knowledge and guidelines about the biology of such important fish like reproduction, food and feeding habits etc. The gonado-somatic index is one of the important parameter of the fish biology which gives the detail idea regarding the fish production, therefore it is under taken for the study from the present area.

Others related works on reproductive biology of fish are, in *Mystus vittatus* (Rao and Sharma, 1984), reproductive cycle and environmental factors in *Mystus vittatus* (Sudha, 2002), reproductive biology of a catfish *Horabagrus bracahysoma* (Molly Kurian and Inasu, 2003) and in *Pimelodus maculates* (Brunopereira *et al.,* 2007) were reported, but none of the above workers have been attempted the study of the Gonadosomatic index of *Mystus cavasius* from this area.

Materials and Methods

The study was conducted during the period of December 2004 to December, 2006 were obtained from fishermen. During the present investigation 205 were males fishes were collected from Bhadra reservoir (13°–42'-00"N latitude and 75°-38°-20"E longitude).The fishes were collected and brought to the laboratory and they were measured accurately for their total length, weight up to the nearest unit. Further fishes were sacrificed after decapitation, length and weight of the fish and weight of the testes were recorded. Gonadosomatic index (GSI) was determined by using the formula of Giese (1959).

$$GSI = \frac{\text{Wet weight of the gonads}}{\text{Wet weight of the fish}} \times 100$$

Results and Discussion

The economic value of any fish depends upon relationship between its length and weight. The ratio of length to the weight of fish is known to be a useful index to demonstrate the well being of the fish (Maceina and Murphy, 1988). In the present investigation GSI range between 0.34 to 1.64 (Table 22.1 and Figure 22.1). The maximum GSI were observed in month of July and minimum in December thus in the

Table 22.1. Month-wise values of GSI of male fish *Mystus cavasius*.

Month	No. of Fish	GSI*
January	15	0.38±0.05
February	25	0.59±0.04
March	16	0.77±0.01
April	19	0.80±0.02
May	20	1.06±0.01
June	15	1.33±0.09
July	14	1.64±0.02
August	20	1.44±0.05
September	15	1.36±0.01
October	18	1.29±0.11
November	18	0.61±0.03
December	10	0.34±0.01

*Values are means of two years (± Standard error).

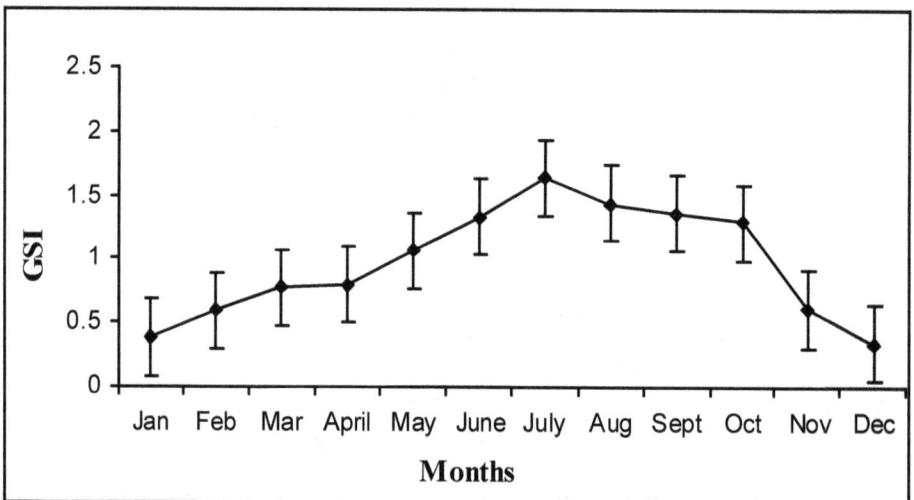

Figure 22.1: Month-wise values of GSI of male fish *Mystus cavasius*.

present investigation the value of GSI indicate pre-spawning season *i.e.* January, February, March and April consisting of immature population. Due to gradual increase in the GSI values indicating prolonged spawning season in the month of May, June, July, August and September indicating population of fully matured individuals which is followed by sudden falls in the GSI values confirming the post spawning season in the month of October, November and December showing the mixed population of matured and spent individuals.

Similarly the maximum the value of GSI were observed in the higher size of length group ranging from 180-220 mm showing the spawning season with population of recorded in the lower size of length group ranging from 130-160 mm (Table 22.2 and Figure 22.2) presenting the immature population and pre-spawning season.

Table 22.2: Size-wise (mm) values of GSI of *Mystus cavasius.*

Length Groups (mm)	Male	GSI*
130-140	27	0.63±0.11
140-150	38	0.51±0.09
150-160	51	0.61±0.03
160-170	20	0.73±0.01
170-180	25	0.73±0.24
180-190	14	0.91±0.07
190-200	10	0.83±0.10
200-210	7	0.99±0.01
210-220	2	1.68±0.06
220-230	3	0.98±0.03
230-240	8	0.58±0.04

*Values are means of two years (± Standard error)

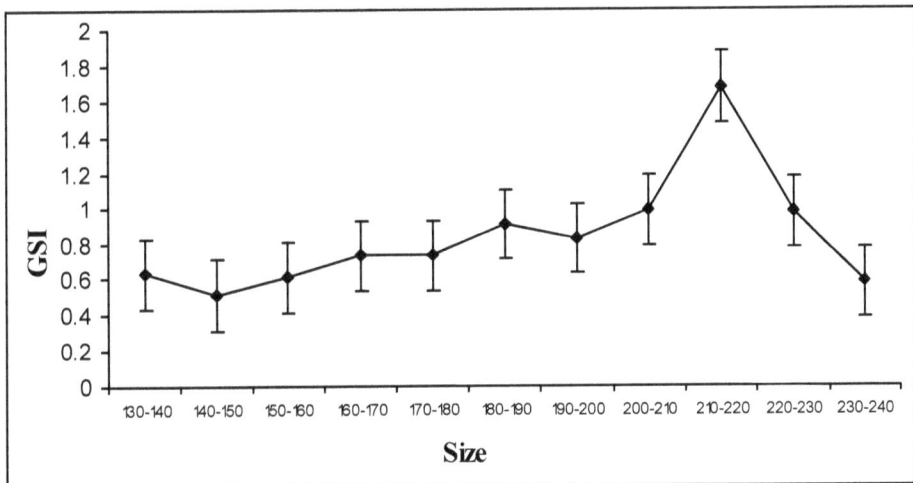

Figure 22.2: Size–wise (mm) values of GSI of *Mystus cavasius.*

The present study can be conclude that there is a gradual increase in the value of GSI up to 180 mm length group and above 180 to 220 mm there is rapid increase in the values presenting immature, mature and spent population of fishes indicating pre-spawning, spawning and post-spawning season in relation to gonadal status and total length of fishes. Thus, value observed in the present study indicating that the GSI is found to be directly correlated with spawning season and having no correlation with a total length of the fish. Similar observations were also reproductive biology of the larvivorous fish *Marcropodus cupanus* (Jocob and Nair, 1983), *Mystus cavasius* (Sharma *et al.,* 1996), grass carp *C. idella* (Shahid and Sheri, 1997) and *Mastacembelus armatus* (Ahirro, 2002).

References

Ahirrao, S.D., 2002. Status of gonads in relation to total length and GSI in fresh water spinney eel, (Pisces) *Mastacembelus armaatus* (Lacepede) from Marathwada Region, Maharashtra. *Aqua. Biol.,* 17(2): 55–57.

Bazzoli, N. and Godinho, H.P., 1991. Reproductive biology of the *Acestrorhynchus lacustris* (Reinhardt, 1874) (Pisces: Characidae) from Tres Marias reservoir, Brazil. *Zoologischer Anzeiger, Munchen.,* 226: 285–297.

Maia, Brunopneira, Ribeiro, S.M.F., P.M. Bizzatto, Vono, Volney and Godinho, Hugo Pereira, 2007. Reproductive activity and recruitment of the yellow-mandi *Pimelodus maculates* (Teleostei: Pimelodidae) in the Igarapava reservoir, Grande river, Southeast Brazil. *Neotropical Icthyology,* 5(2): 147–152.

Giese, A.C., 1959. Comparative physiolgy: Annual reproductive cycle of marine invertebrates. *Ann. Rev. Physiolgy,* 21: 547–560.

Jacob, S.S. and Nair, N.B., 1983. Reproductive biology of the larvivorous fish *Marcropodus cupanus* (Cuv. and Cal). *Proc. Indian Nat. Sci. Acad.,* Part III, 92(2): 159–170.

Kurian, Molly and Inasu, N.D., 2003. Reproductive biology of a catfish *Horabagrus bracahysoma* (Gunther) from inland waters of central Kerala. *J. Inland Fish. Soc. India,* 35 (1): 1–7.

Maceina, M.J. and Murphy, B.R., 1988. Variation in the weight-to-length relationship among Florida and Northern large mouth bass and their intraspecific F1 hybrid. *Trans. Am. Fish. Soc.,* 117: 232–237.

Shahid, M. and Sheri, A.B., 1997. Relationship among ovary weight, liver weight and body weight in female grass carp *C. idella. J. Aqua. Trop.,* 12: 255–259.

Sharma, S.V., Jaya Raju, P.B. and Rao, T.J.S. Alankara, 1996. Reproductive biology of tropical fresh water catfish *Mystus cavasius* (Siluriformes: Bagridae) from Guntur, India. *J. Aqua. Biol.,* 11(1 and 2): 33–40.

Sudha, H.R., 2002a. Reproductive cycle and environmental factors in *Mystus vittatus* (Bloch). *Uttar Pradesh J. Zool.,* 22(1): 25–32.

Thapliyal, Anju and Dobriyal, Anoop K., 2003. Maturation and spawning biology of a hill stream catfish *Pseudecheneis* (McClelland) from the Garhwal Himalaya, Uttaranchal. *Ad. Bios.,* 22(1): 9–22.

2013, Glimpses of Animal Biodiversity

Pages **212–218**

Editor: **Dr. K. Muthuchelian,** *Vice Chancellor, Periyar University, Salem*

Published by: **Daya Publishing House, NEW DELHI**

Chapter 23

A Study on Interactions of Insects with *Crotalaria retusa* L. (Fabaceae)

N. Govinda Rao, D. Jeevan Ram and A.J. Solomon Raju

*Department of Environmental Sciences,
Andhra University, Visakhapatnam – 530 003*

ABSTRACT

C. retusa is a seasonal shrub. In *C. retusa,* the flowers conceal the stamens and stigma in keel petals which in turn escorted by wing petals. The specialist pollinators while collecting nectar depress the keel and wing petals in order to reach the nectar deeply seated at the corolla base. In consequence, the keel petals release the stamens and stigma violently due to which pollen from the dehiscent anthers is shed unto the ventral side of the specialist pollinator. The pollinator at the same time gains contact with the stigma and effects pollination. The pollinator bees are exclusively *Xylocopa latipes* and *X. pubescens.* The flowers also attract three Nymphalid butterfly species, *Danaus chrysippus, Euploea core* and *Tirumala limniace, D. chrysippus* and *E. core* visit the mature buds and flowers to collect nectar while *T. limniace* in huge aggregations collect sap from the stem and leaf petioles before flowering and after fruit set. *D. chrysippus* and *E. core* are specialist butterflies to collect nectar which is known for its pyrrolizidine alkaloid content. *T. limniace* is a specialist feeder of the sap which is also known for its alkaloid content. The alkaloids enable these butterflies to protect themselves from their predators and synthesize their courtship pheromone.

Keywords*: Crotalaria retusa, Butterflies, Carpenter bees, Alkaloids.*

Introduction

The genus *Crotalaria* is a classical example for its specialized papilionaceous flowers adapted for specialist pollinators. Majinda *et al.* (2001) reported that *C. pallida, C. lanceolata* and *C. retusa* are adapted for pollination exclusively by large carpenter bees, *Xylocopa frontalis* and *X. grisescens.* Solomon Raju (2007) reported that *C. laburnifolia* and *C. verrucosa* are exclusively pollinated by *Xylocopa latipes* and *X. pubescens.* In the present study, the sap collection activity of butterflies and nectar collection activity of bees and butterflies on *Crotalaria retusa* with reference to alkaloid source and pollination role were explained in detail.

Materials and Methods

Crotalaria retusa occurring at the Seshachalam Hills of southern Eastern Ghats of Andhra Pradesh was used for the study during the summer season of 2009. Twenty five tagged mature buds were followed for recording the time of anthesis and anther dehiscence. The details of flower morphology such as flower sex, shape, size, colour, odour, sepals, petals, stamens and ovary were described. Twenty mature but undehisced anthers was collected from different plants and placed in a Petri dish. Later, each time a single anther was taken out and placed on a clean microscope slide (75 x 25 mm) and dabbed with a needle in a drop of lactophenol-aniline-blue. The anther tissue was then observed under the microscope for pollen, if any, and if pollen grains were not there, the tissue was removed from the slide. The pollen mass was drawn into a band, and the total number of pollen grains was counted under a compound microscope (40x objective, 10x eye piece). This procedure was followed for counting the number of pollen grains in each anther collected. Based on these counts, the mean number of pollen produced per anther was determined. The mean pollen output per anther was multiplied by the number of anthers in the flower for obtaining the mean number of pollen grains per flower. The characteristics of pollen grains were also recorded. The average volume of nectar per flower was determined and expressed in μl; for this twenty fresh flowers were used. The flowers used for this purpose were bagged at mature bud stage, opened after anthesis and squeezed nectar into micropipette for measuring the volume of nectar. The stigma receptivity was observed visually and by H_2O_2 test as given in Dafni *et al.* (2005). Regular observations were made on the sap collecting butterflies and nectar feeding butterflies and bees. They were observed for their foraging behaviour such as mode of approach, landing, probing behaviour for nectar collection, contact with essential organs to result in pollination, inter-plant foraging activity in terms of cross-pollination, etc. The foraging visits of each butterfly species for the entire day on the selected number of flowers were recorded to know the foraging trend from morning to evening.

Results

It is an under-shrub with glabrous or appressed-pubescent branches. Leaf is simple, oblanceolate-oblong, retuse, glabrous above, silky pubescent below, petiole very short, stipules subulate. It flowers during November-February with more flowering rate between mid-December and late-January. The inflorescence is an erect, terminal raceme producing an average number of 25 flowers which anthese acropetally. The flowers are medium-sized, yellow, showy, bisexual and zygomorphic with five 13 mm long glabrous sepals united to form a small tube. The corolla is papilionaceous, golden-yellow with purple tinge; each petal is 25 mm long. The stamens are ten, five small and five long with monadelphous condition; the anthers are 2-celled, introrse, and basifixed. The ovary with an average of 18 ovules are arranged on marginal placentation and terminated with a long style and a small hairy stigma. The seeds become loose in the pod as they mature, and rattle when the pod is shaken. Fruit is glabrous, 2.5-4.8 cm long, linear-oblong, 15-20 seeded, stipitate. Pods are inflated, dark brown to black at maturity, 3-4 cm long, seeds tan to black.

The Nymphalid butterfly, *Tirumala limniace* collected sap from the stem and leaf petioles consistently. More than 800 individuals of this butterfly were found to aggregate on *C. retusa* plants for the collection of sap. This butterfly utilized the plants prior to flowering and after fruit set for sap collection. Such plants subsequently withered. It never visited the mature buds and flowers for nectar collection and also it has not utilized the plant as a larval host.

The flowers open during 1400-1700 h with more anthesis at 1500-1600 h. Anther dehiscence in the long stamens takes place ca. 36 hours before anthesis while that in the short stamens occurs two hours ahead of anthesis. In both types of stamens, anther dehiscence occurs by longitudinal slits. The pollen production per anther averages 9,028 in the long stamens and 3,310 in the smaller ones. The pollen-ovule ratio is 3427:1. The pollen grains are light yellow, dry and powdery. The stigma attains receptivity 30 min after anther dehiscence in the short stamens. The stigma in receptive phase is wet and shiny and remains so day-long. The floral buds begin nectar secretion ca. 4 hours before anthesis and continue secretion until ca. 2 hours after anthesis. A flower secretes 9.35 ± 0.89 µl of nectar. The floral nectar is accessible by butterflies with long proboscides and by other insects which have the ability to depress carinal keel petals with their body weight and forcible probing behaviour. The carpenter bees, *Xylocopa latipes* and *X. pubescens* were the only bee species visiting the flowers for nectar. They depressed the keel and wing petals to access the nectar situated at the corolla base. This probing behavior resulted in the violent release of stamens and stigma from the keel petals; at the same time self- or cross-pollination occurs. Therefore, these two carpenter bees were the efficient and exclusive pollinators.

The Nymphalid butterflies, *Danaus chrysippus* and *Euploea core* collected nectar with their long proboscis from the mature buds and open flowers for nectar consistently throughout the flowering season. They approached the mature bud in upright position and penetrated the sealed corolla with their proboscis to imbibe nectar; they accessed the nectar in open flowers with great ease. They collected nectar throughout the day with more activity during afternoon period. They visited mature buds in preference to open flowers of the previous day during forenoon period and concentrated on open flowers during the afternoon period (Figures 23.1 and 23.2). These butterfly species were the exclusive foragers during the forenoon period while carpenter bees collected nectar during afternoon period only along with the butterfly species.

Discussion

The genus *Crotalaria* is a classical example for its specialized papilionaceous flowers adapted for specialist pollinators. The flowers conceal the stamens and stigma

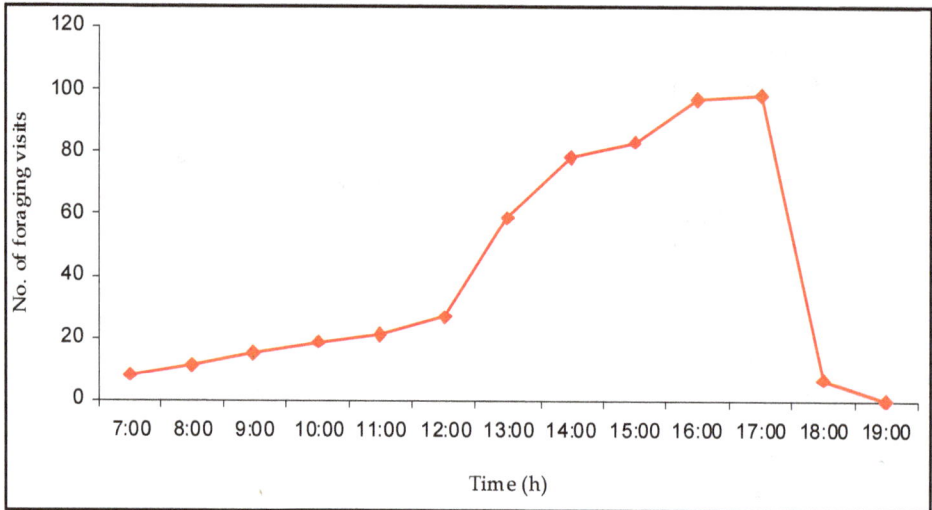

Figure 23.1: Nectar foraging activity of *Danaus chrysippus* and *Crotalaria retusa*.

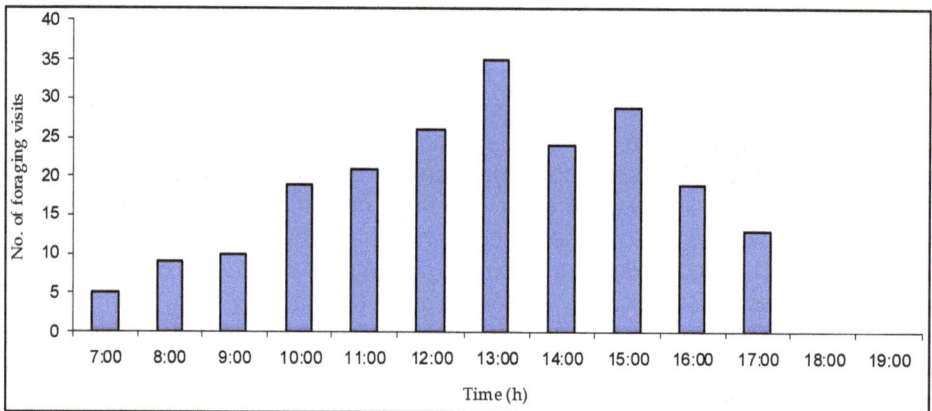

Figure 23.2: Nectar foraging activity of *Euploea core* on *Crotalaria retusa*.

in keel petals which in turn are escorted by wing petals. The specialist pollinators while collecting nectar depress the keel and wing petals in order to reach the nectar deeply seated at the corolla base. In consequence, the keel petals release the stamens and stigma violently due to which pollen from the dehiscent anthers is shed unto the ventral side of the specialist pollinator. The pollinator at the same time gains contact with the stigma and effects pollination. Majinda *et al.* (2001) reported that *C. pallida*, *C. lanceolata* and *C. retusa* are adapted for pollination exclusively by large carpenter bees, *Xylocopa frontalis* and *X. grisescens*. Solomon Raju (2007) reported that *C. laburnifolia* and *C. verrucosa* are exclusively pollinated by *Xylocopa latipes* and *X. pubescens*. In *C. retusa* also, the same species of carpenter bees effect pollination.

C. *retusa* at the study sites also attracts three Nymphalid butterfly species, *Danaus chrysippus*, *Euploea core*, *Tirumala limniace*. The first two species visit the mature buds and flowers to collect nectar while *T. limniace* in huge aggregations collect sap from the stem and leaf petioles before flowering and after fruit set. *D. chrysippus* and *E. core* collect nectar mostly from mature buds than flowers suggesting that mature buds are more attractive and the reason for this may be due to the presence of full volume of fresh nectar while flowers may not contain full volume of nectar and the nectar is also relatively not fresh. These butterfly species are capable to probe the mature buds with their proboscis and penetrate the corolla to imbibe nectar. Bhuyan *et al.* (1999) reported that *Danaus genutia* exhibits such a probing behaviour to collect nectar from mature buds and flowers of *C. juncea*. These authors also mentioned that this butterfly spends more time at mature buds to probe and collect full volume of fresh nectar.

The genus *Crotalaria* is known to produce mainly pyrrolizidine alkaloids (Roeder *et al.*, 1993). In the arctiid moth, *Utetheisa ornatrix*, males sequester pyrrolizidine alkaloids as a larva from *Crotalaria* spp., advertise the magnitude of their alkaloidal gift in courtship, by way of a pheromone and transmit them as a nuptial gift to females during courtship. The females bestow this gift on her eggs, together with a fraction of her own sequestered alkaloids. All stages of development, including the eggs as a consequence of their biparental endowment, are protected by the alkaloid against predation (Eisner and Meinwald 1987; Eisner and Eisner 1991; Dussourd *et al.*, 1991). Jones and Blum (1983) reported that the pyrrolizidine alkaloids play a key role in host-plant selection and as a sex pheromone in certain danaid butterflies. These butterflies apply from their proboscides a fluid capable of dissolving alkaloids and then re-imbibe it. Both sexes of these butterflies store alkaloids apparently for defense (Edgar *et al.*, 1976; 1979; Rothschild *et al.*, 1979, Conner *et al.*, 1981). Owen (1971) reported that danaine butterflies use withered and damaged plants and floral nectar as sources of alkaloids. They also scratch and suck at living plants. Johnston and Johntson (1980) reported that danaine butterflies scratch leaves of *Crotalaria retusa* and imbibe the sap which oozes out. In the present study, danaine butterfly *D. chrysippus* now placed in Nymphalidae family and *E. core* are the only butterfly species which are attracted to the mature buds and flowers of *C. retusa*. This observation conforms to the finding by Nicolson (2007) that pyrrolizidine alkaloids in nectar are inhibitory to generalist-feeding butterflies but attractive to specialist feeders. Therefore, *D. chrysippus* and *E. core* are specialist feeders of the nectar of *C. retusa*. These butterflies tend to stay longer at the mature buds and flowers to imbibe nectar; it could be partly due to intoxication with narcotic substances present in the nectar; this leads to "sluggish" behaviour of the butterflies. Such behaviour increases the time spent at buds and flowers and the chances of successful pollination; the latter happens only if its proboscis carries enough pollen grains to pollinate the flowers (Nicolson 2007). *D. chrysippus* and *E. core* use the pyrrolizidine alkaloids acquired from *C. retusa* to protect from their predators and synthesize the courtship pheromone; maternal and parental contributions of alkaloids play an important role to protect the most vulnerable stage, the egg (Meinwald 1990). These butterfly species and also *D. plexippus* when equipped with these alkaloids have been experimentally proved to be unpalatable to their predators by Edgar *et al.* (1976).

In the present study, the sap feeding activity of *T. limniace* confined to the stem and leaf petioles of *C. retusa* suggests that it is not a specialist feeder of its nectar and hence it totally avoids mature buds and flowers though they are sources of pyrrolizidine alkaloids. However, this butterfly is a specialist feeder of the sap of *C. retusa*; it also seems to have inherent ability to sense and collect sap before and after flowering season. Such sap collection activity suggests that the plant might be rich in alkaloid content prior to and after flowering. The butterfly collects the sap by making surface scratches on the leaf and by slightly damaging the tissue of leaf petioles. Boppre (1983) reported similar sap collection behaviour by *Tirumala petiverana* on the Boraginaceae member, *Heliotropium pectinatum* in East Africa. Mathew and Anto (2007) reported that *T. limniace* uses the pyrrolizidine alkaloids to deter its predators, the garden lizards. Therefore, *T. limniace* uses *C. retusa* is an important source of alkaloids and hence this butterfly concentrates and congregates in the patches of this plant prior to flowering and during pod set stage. The plants fed before flowering gradually wither while those fed during pod set stage result in the abortion of developing pods. The flowering occurs only in those plants that have not been used by *T. limniaceae*. Further, it is important to mention that it does not use *C. retusa* as a larval host plant.

Hall (2005) stated that *Crotalaria* species are toxic to livestock due to the presence of pyrrolizidine alkaloids and can potentially be fatal. *C. lanceolata, C. pallida* and *C. spectabilis* introduced in the United States of America have been considered to be fatal to livestock. In this study, the local people reported that *C. retusa* is toxic to cattle and it is difficult for them to monitor their cattle when allowed them to graze on grass in the forest areas. In this context, the butterfly *T. limniace* plays a beneficial role by causing withering of plants before flowering and causing the abortion of the developing pods; this shows severe impact on the population size and also on the success of sexual reproduction in *C. retusa*.

References

Bhuyan, M., Kataki, D., Deka, M. and Bhattacharya, P., 1999. Nectar host plant selection and floral probing by the Indian butterfly *Danaus genutia* (Nymphalidae). *J. Res. Lepidoptera,* 38: 79–84.

Boppre, M., 1983. Adult Lepidoptera "feeding" at withered *Heliotropium* plants (Boraginaceae) in East Africa. *Ecol. Entomol.*, 6: 449–452.

Conner, W.E., Eisner, T., Vander Meer, R.K., Guerrero, A., Ghitingelli, D. and Meinwald, J., 1981. Precopulatory sexual interaction in an arctiid moth (*Utetheisa ornatrix*): Role of a pheromone derived from dietary alkaloids. *Behavioural Ecolo. and Sociobiol.*, 9: 227–235.

Dafni, A., Kevan, P.G. and Husband, B.C., 2005. *Practical Pollination Biology.* Enviroquest Ltd., Cambridge.

Dussourd, D.E., Harvis, C.A., Meinwald, J. and Eisner, T., 1991. Defense mechanisms of arthropods. 107. Pheromonal advertisement of a nuptial gift by a male moth (*Utetheisa ornatrix*). *Proc. Natl. Acad. Sci., USA,* 88: 9224–9227.

Edgar, J.A., Boppre, M. and Schneider, D., 1979. Pyrrolizidine alkaloid storage in African and Australian danaid butterflies. *Experientia,* 35: 1447–1448.

Edgar, J.A., Cockrum, P.A. and Frahn, J.L., 1976. Pyrrolizidine alkaloids in *Danaus plexippus* L. and *Danaus chrysippus* L. *Experientia (Basel),* 32: 1535–1537.

Eisner, T. and Eisner, M., 1991. Unpalatability of the pyrrolizidine alkaloid-containing moth *Utetheisa ornatrix,* and its larva, to wolf spiders. *Psyche,* 98: 111–118.

Eisner, T. and Meinwald, J. 1987. Alkaloid-derived pheromones and sexual selection in Lepidoptera. In: *Pheromone Biochemistry,* (Eds.) G.D. Prestwich and G.J. Blomquist. Academic Press, Florida, pp. 251–269.

Hall, D.W., 2005. Bella moth, rattlebox moth, Inornate moth or Calico moth, *Utetheisa ornatrix* (Linnaeus) (Insecta: Lepidoptera; Arctiidae: Arctiinae). Website at: http://creatures.ifas.ufl.edu.

Johnston, G. and Johnston, B., 1980. *This is Hong Kong: Butterflies.* Hong Kong Government Publication.

Jones, T.H. and Blum, M.S., 1983. Arthropod alkaloids: Distribution, functions and chemistry. In: *Alkaloids: Chemical and Biological Perspectives,* (Ed.) S.W. Pelletier. Wiley, New York, pp. 34–84.

Majinda, R.R.T., Abegaz, B.M., Bezabih, M., Ngadjui, B.T., Wanjala, C.C.W., Mdee, L.K., Bojase, G., Silayo, A. Masesane, I. and Yeboah, S.O., 2001. Recent results from natural product research at the University of Botswana. *Pure Appl. Chem.,* 73: 1197–1208.

Mathew, G. and Anto, M., 2007. *In situ* conservation of butterflies through establishment of butterfly gardens: A case study at Peechi, Kerala, India. *Curr. Sci.,* 93: 337–347.

Meinwald, J., 1990. Alkaloids and isoprenoids as defensive and signaling agents among insects. *Pure and App. Chem.,* 62: 1325–1328.

Nicolson, S.W., 2007. Nectar consumers. In: *Nectaries and Nectar,* (Eds.) S. Nicolson *et al.* Springer, Netherlands, pp. 289–342.

Owen, D.F., 1971. *Tropical Butterflies.* Clarendon Press, Oxford.

Roeder, E., Sarg, T., El-Dahmy, S. and Ghani, A.A., 1993. Pyrrolizidine alkaloids from *Crotalaria aegyptiaca. Phytochemistry,* 34: 252–253.

Rothschild, M., Alpin, R.T., Cockrum, P.A., Edgar, J.A., Fairweather, P., 1979. Pyrrolizidine alkaloids in Arctiid moths (Lepidoptera) with a discussion on host plant relationships and the role of these secondary plant substances in the Arctiidae. *Biol. J. Linnean Soc.,* 12: 305–326.

Solomon Raju, A.J., 2007. *Indian Plant Reproductive Ecology and Biodiversity.* Today and Tomorrow's Printers and Publishers, New Delhi.

Chapter 24

Leaf Webber-cum-Fruit Borer, *Pempelia morosalis*: One of the Major Threats in Biodiesel Production of Jatropha

B. Usha, G. Jayalaxmi, R.N. Ganguli and D.J. Pophaly

*Department of Entomology, College of Agriculture,
Indira Gandhi Agricultural University, Raipur, Chhattisgarh*

ABSTRACT

The present studies were conducted in *Jatropha curcas* at Raipur, Chhattisgarh, during 2007-08. The webber cum fruit borer, *Pempelia morosalis* (Saalm uller) (Lepidoptera: Pyralidae) is one of the major threats in the bio-diesel production of Jatropha as it causes damage to almost all parts of the plant. The dark green active larvae web the leaves, apical stems and inflorescence feeding voraciously which affects the fruit set very badly. The larvae showed its activity from October to June. Maximum (103.98 larvae/plant) was recorded during 2nd fortnight of October. The larvae were also seen to bore inside the fruits which deteriorated the quantity and quality of oil production of seeds.

Introduction

Jatropha curcas is a well known species yielding bio-diesel for which it is valued for the future. An additive advantage of the species is that it can be grown on waste and barren lands. The "curcas oil" apart from being used as supplement of diesel has multiparous uses in soap, cosmetic and candle industry. Among the various threats faced by the species, insect pests are the major ones which deteriorate the plant growth, vigour and oil production adversely. Among the various major and minor insect pests, the leaf webber cum fruit borer, *Pempelia morosalis* (Saalm uller) is one of the major one damaging almost all parts of the plant. It webs the leaves, apical stem, inflorescence and also bore into the fruits, thus affecting the seed quality and oil production.

Materials and Methods

The present investigation was carried during September 2007 to July, 2008. The experiment was conducted in Randomized Block Design with 47 provenances each replicated three times.

Experimental Details

☆ Design : RBD
☆ Replication : Three
☆ Number of provenances : 47
☆ Plot size : 90 x 40 m^2
☆ Date of planting : 27.02.2005

The data was analyzed in RBD considering two factors *i.e.* the various types and number of larvae of *P. morosalis* infesting Jatropha and the number of natural enemies, applying square root transformations. For recording observations, the whole experimental field was divided into 35 blocks, each block having 9 plants. The observations was recorded at 15 days interval for number of larvae from each block on 3 randomly selected plants from each provenance and replicated thrice. Data was analyzed in randomized block design using appropriate transformation.

Description of the Insect

☆ *Eggs*: The eggs are laid in clusters mostly at the basal part of inflorescence, which are covered with a thin creamy white coloured film. Single female laid about 20-25 eggs, in masses.

☆ *Larvae*: The larvae were dark green to brownish green in colour. The newly hatched larvae measured 0.2 cm in length and 0.10 cm in width.

☆ The first instar larvae were green with brown head.

☆ They were very active and fed voraciously on young leaves and inflorescence of the Jatropha plants.

☆ Full grown larvae measured 1.90 cm in length and 0.30 cm in width with dark brownish green in colour. At flowering stage they bore in to peduncle and fruits which developed galleries made of silk and frass.

☆ The larvae bored into the fruits throwing out faecal matter. There were five larval instars, and the average duration of larval period was observed to be 29.33 days.

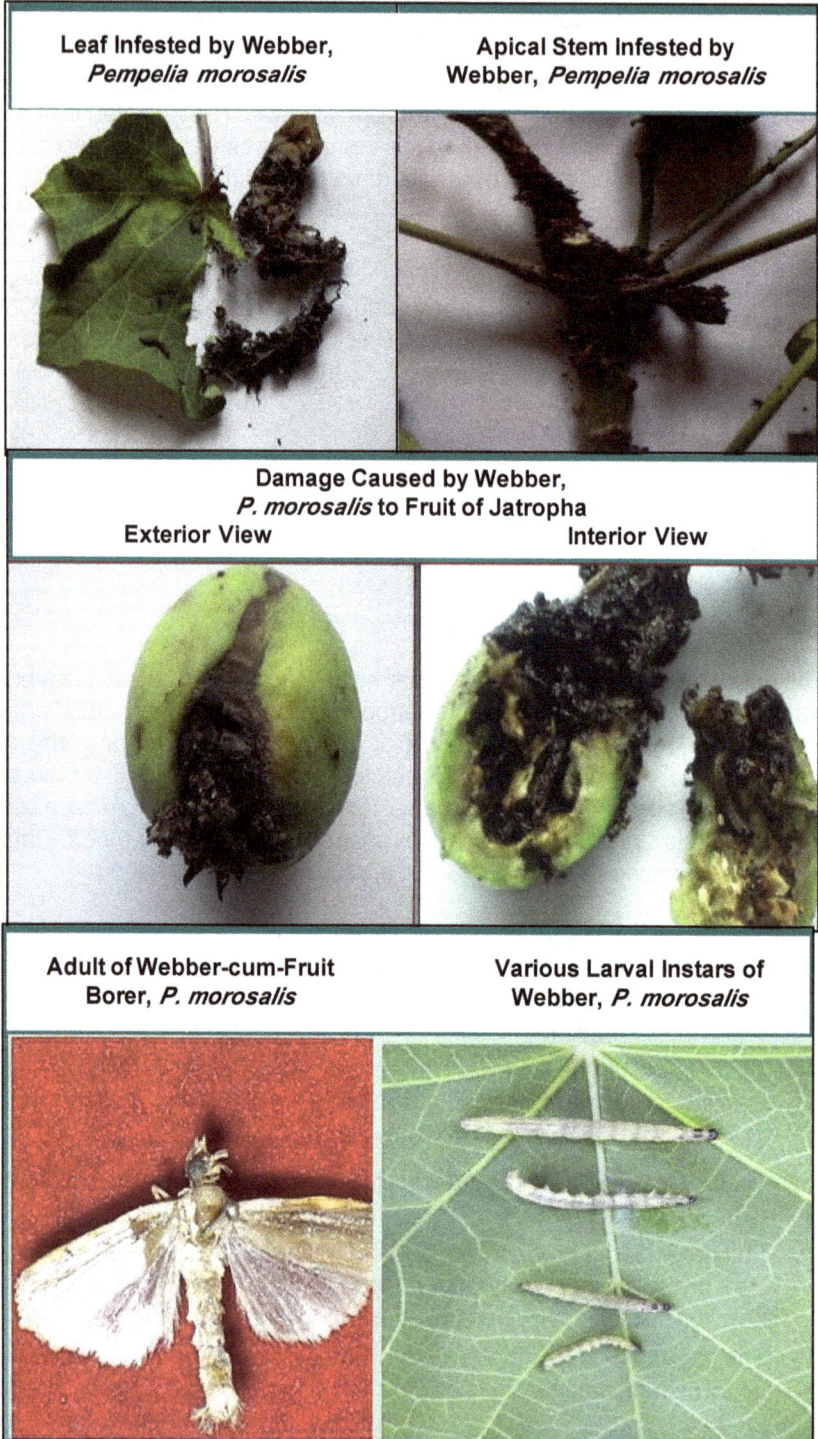

| Leaf Infested by Webber, *Pempelia morosalis* | Apical Stem Infested by Webber, *Pempelia morosalis* |

Damage Caused by Webber, *P. morosalis* to Fruit of Jatropha

Exterior View   Interior View

| Adult of Webber-cum-Fruit Borer, *P. morosalis* | Various Larval Instars of Webber, *P. morosalis* |

Figure 24.1

Results and Discussion

The maximum population of larvae of webber, *P. morosalis* (103.98 larvae/plant) was recorded on IInd fortnight of October and the minimum population of (34.80 larvae/plant) were observed during the IInd fortnight of November. No population of webber larvae was observed in the month of September (Table 24.1) in season I. In season II, the maximum and minimum population was recorded 1st fortnight of April (91.75 larvae/plant) and 1st fortnight of June (2.56 larvae/plant), respectively. The larvae were found damaging almost all plant parts namely, leaves, apical stem, inflorescence and also fruits. The inflorescence attacked by webber did not show any fruit set. *P. morosalis* is emerging as a major problem in Tamil Nadu. Monoharan *et al.* (2006) among the listed various pests, has stated that it is specific to Jatropha and reported to affect forest species like *Desmodium gangeticum, Flemingia* sp. and *Uraia lagopides.* Soman *et al.* (2006) and Tamrakar *et al.* (2007) also reported webber as a pest of Jatropha in Chhattisgarh.

Table 24.1: List of provenances of *Jatropha curcas.*

Sl. No.	Name of Provenances	Sl. No.	Name of provenances
1.	Sagar-1	25.	J&K Set 2
2.	RJ 117 (A)	26.	Jagdalpur
3.	Dehradoon	27.	Kalyanpur
4.	Barbuspur	28.	APOS-2001
5.	Pant J and K Set 2	29.	RJ 117 (B)
6.	J and K Set 1	30.	TNMC-7
7.	Jabalpur	31.	ANOS-201
8.	J and K Set 1	32.	PKVJ-SJ-1
9.	NRCAF-13	33.	Pendra Road
10.	Baikunthpur	34.	PKVJ-DHW-1
11.	TNMC-5	35.	NRCAF-14
12.	Mandeshwar	36.	Balodabazar
13.	Pant J and K Set 1	37.	Kot
14.	PKVJ-MKV-1	38.	Tukupoms
15.	AMOS-201	39.	Taraipur
16.	Bawal	40.	Kilkila
17.	NRCAF-15	41.	Chandrapur
18.	PKVJ-AKT-1	42.	Mahanrpur
19.	Indore-I	43.	Surajpur
20.	Korba	44.	Sonhat
21.	Chandka	45.	Saheltarai
22.	TFRI-1	46.	Churmundra
23.	Barmunda	47.	Keshipur
24.	NRCAF-18		

Table 24.2: Mean fortnightly population of webber, *P.morosalis* in *J. curcas.*

Sl.No.	Months	Mean No. of Larvae/Plant
	Season-I	
1.	Sept. 1st FN	0.000
2.	Sept. 2nd FN	0.000
3.	Oct. 1st FN	0.000
4.	Oct. 2nd FN	103.982
5.	Nov. 1st FN	78.658
6.	Nov. 2nd FN	34.801
	Season-II	
7.	April 1st FN	91.753
8.	April 2nd FN	69.610
9.	May 1st FN	46.651
10.	May 2nd FN	14.327
11.	June 1st FN	2.553
12.	June 2nd FN	0.000

Conclusion

Thus from the above studies we can conclude that the leaf webber cum fruit borer, *P.morosalis* is emerging as a big threat not only in Chhattisgarh but also in many other parts of India. Since, the pest attacks almost all plant parts causing direct effect on the growth, flowering and fruit set ultimately reducing the oil production considerably.

References

Manoharan, T., Ambika, S., Natrajan, N. and Senguttuvan, K., 2006. Emerging pest status of *Jatropha curcas* (L) in South India. *Indian J. of Agroforestry,* 8(66): 79.

Soman, D., Ganguli, J.L. and Ganguli, R.N., 2006. Biodiversity of insect pests of multitier agroforestry system. In: Paper presented in *National Conference on Forest Biodiversity, Resources : Exploitation, Conservation and Management,* held at CBFS, Madurai Kamraj University, Madurai form 21–22 March. pp. 113–114.

Tamrakar, E., Ganguli, J.L., Soman, D. and Ganguli, R.N., 2007. Studies on pest succession on Jatropha (*Jatropha curcas*) under multitier agroforestry system. In: Paper presented in *National Symposium on Sustainable Pest Management for Safer Environment,* held at OUAT, Bhubaneshwar from 6–7 Dec.

2013, Glimpses of Animal Biodiversity   *Pages* **227–226**
Editor: **Dr. K. Muthuchelian,** *Vice Chancellor, Periyar University, Salem*
Published by: **Daya Publishing House, NEW DELHI**

Chapter 25

Protective Effect of 'Bactrocure' against HCH– Pesticide in Groundnut Soil

Doris M. Singh* and Kiran Kolkar
Department of Botany, Karnatak Science College, Dharwad

ABSTRACT

Twenty earthen pots of 12 kg capacity were filled with 10 kg each of garden soil. Six treatments were maintained in duplicate.1. Control; 2. 100ppm HCH; 3. 100ppm + Manure; 4. 100ppm HCH + Bactrocure; 5. Manure; 6. Bactrocure. Study with groundnut plants has shown that Bactrocure containing *Pseudomonas tralucida* can be directly deployed in the soil environment to remove the residue. Such a curative treatment of the soil removes phyto-toxicity. As noticed in 30 days the Bactrocure degraded HCH by 35.1 per cent and by 90 days another 28.6 per cent. It is also corrects the physiological processes in groundnut. As seen in chlorophyll synthesis which was almost 100 per cent. Nodulation in the plant was highest in Bactrocure 45-50 N/plant and equal to manure treatment alone when treated with 100ppm HCH 30 N/plant. Yield g/pot was 44.1 with Bactrocure and it doubles the yield of HCH treatment 15.5 and 8.0 respectively. Oil content showed a similar trend. Therefore in this appraisal sustainable development of soil by residue degradation has been reviewed.

Keywords: Bactrocure, Pseudomonas tralucida, HCH, Phytotoxicity, Physiology.

---

* Corresponding Author: E-mail: dorissingh@ gmail.com

Introduction

With the increasing number of pesticides used, there is simultaneously an increase in pesticide consumption. The toxicological implications of the extensive use of pesticide in agriculture are consequently of great relevance to the causes of ill-health. This persistent chemical causes, harm to the activity and population of soil microorganisms in leguminous soils affecting the physiological balance of the plants (Mahendranath 1986). Since the classical demonstration of microbial break down of the hormone pesticide 2, 4-D (Audus 1951), it has become established that the major, or frequently the only means of degradation for several pesticides in the environment is mainly mediated by microorganisms. Therefore sustainable development of soil by residue degradation can be achieved as in the use of 'Bactrocure' a microbial preparation which has proved efficient in reversing the toxic effect and giving improved growth of plant by protecting against phytotoxicity, inhibitory activity of soil enzyme resulting in better yield.

Materials and Methods

Pot Culture Study

Twenty earthen pots of 12 kg capacity were filled with 10 kg each of garden soil. Six treatments were maintained in duplicate as follows: 1. Control; 2.100ppm HCH; 3. 100ppm + Manure; 4. 100ppm HCH + Bactrocure; 5. Manure; 6. Bactrocure. Bactrocure 40mg/10kg; Horse dung compost 10g/10kg and α–HCH dissolved in acetone and added to get 50 and 100ppm concentration were mixed to a small quantity of soil. This sample was then added to appropriate soil and thoroughly mixed to get uniform concentration. Water was sprinkled to attain 60 per cent water holding capacity and maintained at that level throughout the course of the experiment. After harvesting (90 days) the weight of the pods and length of the shoot, roots were recorded.

Estimation of Chlorophyll

Chlorophyll was estimated according to AOAC, (1980).

Extraction and Analysis of Pesticide Residue in Soil

The method of Visweswariah *et al.* (1973) was followed to determine HCH residue in treated soil.

Estimation of Oil Content

Oil content in groundnut sample was attained using the soxhlet method.

Estimation of Yield and Nodules

Nodules were counted, while yield was obtained using digital balance.

Bactrocure

This is a biological product developed at C.F.T.R.I for abatement of pesticide residue in soil. Essentially the preparation contains Pseudomonas capable of degrading HCH mixed in lignite powder at neutral pH.

Results and Discussion

Organo-chlorine insecticides such as HCH are widely used in agricultural production. Different reports from different research organizations spread over the country reveal, HCH residues in almost all environment and biological sciences (Kishna Murti, 1984).

Addition of Bactrocure showed boosting effect in the parameters studied. HCH or manure treated soils caused chlorosis of leaves in plants originating from it. A maximum reduction of 42 per cent was seen due to 100ppm HCH in soil and around 34 per cent with manure. But Bactrocure treated HCH contaminated soil supports better growth of plants as in chlorophyll synthesis which was almost 100 per cent. With the use of HCH 100ppm–15 nodules/plant, manure -22 nodules/plant and manure + HCH-20 nodules/plant. There was a significant reduction in nodule count while with Bactrocure 45-50 nodules/plant was obtained. Anusuya (1988) has described reduction in wet weight of nodules and leg haemoglobin content after foliar spray with antibiotic cyclohexamide in groundnut. Presence of HCH or manure individually did not reduce the oil content. But Bactrocure definitely caused improvement in the oil content. The significant improvement was seen in manure HCH Bactrocure combination. The oil content increased from 32 per cent to 37 per cent. The application if HCH to soil gave the least number of pods/plant; number of seeds/plant and number of seeds/pod was the least 4.4, 7.0, and 1.6 respectively. While with the application of Bactrocure boosted the yield to 8.7 pods/plant; 12.2 numbers of seeds/plant and the number of seeds/pod was the same. With the addition of Bactrocure to manure treated soil gave the best results 13.9 pods/plant; 22.4 number of seeds/plant and 1.6 number of seeds/pod. This clearly indicates a curative treatment of the soil brings about an over all boosting effect on the processes, removes phytotoxicity and corrects the physiological processes in groundnut, and ultimately prevents HCH accumulation in the plant parts. Sethunathan *et al.* (1982) suggested the potential of microorganisms in residue abatement. However, the study reveals achievement of residue abatement deploying bacteria.

References

Anusuya, D., 1988. *Pesticides,* 22 (3): 17–19.

Audus, J.L., 1951. *Plant Soil,* 3: 170.

Kishna Murti, C.R., 1984. Indian National Science Academy, Bahadur Shah Zafar Marg, New Delhi.

Mahendranath, T.R., Gupta, A.K. and Varshney, M.L., 1986. *Pesticide,* 20(7): 36–38.

Sethunathan, N., Adhya, T.K. and Raghu, K., 1982. *Biodegradation of Pesticides,* (Eds.) F. Mastumura and C.R. Krishna Murti. Plenum Press, New York.

2013, Glimpses of Animal Biodiversity   *Pages* **227–232**
Editor: **Dr. K. Muthuchelian,** *Vice Chancellor, Periyar University, Salem*
Published by: **Daya Publishing House, NEW DELHI**

Chapter 26

Inventory of Ethnomedicinal Plants in the Dry Pond Vettangudi Birds Sanctuary, Sivaganga District, Tamil Nadu

T. Arun Raja and D. Kannan*
Department of Botany,
Thiagarajar College, Madurai – 625 009

ABSTRACT

Pond habitat is a unique ecosystem encompassing varying organism, including medicinal plants. Medicinal plants are greatly utilized for various purposes including codified, uncodified system of medicine, local collection, trade and commerce. Documentation on medicinal plants and their inventory are importantly considered. Biodiversity nature and the ecological functioning of ponds are greatly varying depends with the associated environmental feature and also the utilization and management of ponds. Medicinal plants survey has been done different landscape; works on such studies are scanty. The present study envisaged the herbal plants diversity collected from the birds' sanctuary pond and the plants utilization for different ailments by the local community of the pond area.

Keywords*: Bird sanctuary, Vettangudi pond, Dry pond vegetation, Wetland ecosystem.*

* Corresponding Author: E-mail: dekan_c@rediffmail.com

Introduction

India harbours about 15 per cent (3000–3500) medicinal plants, out of 20,000 species of medicinal plants in the world. About 90 per cent of these are found growing wild in different climatic regions of the country (Singh, 1997). Medicinal plants have great economic value in the Indian subcontinent (Parekh *et al.,* 2005). The Indian region has rich cultural diversity and so, ethnobotanical heritage is greatly valued (Jain, 1991). More than 7,500 medicinal plant species are used for healthcare by various ethnic communities in India (Arora, 1997). The study of ethnobotany assumes great importance in enhancing our knowledge.

Each and every local people use certain plants as medicine. They are using specific remedies prepared from selected plants for the treatment of particular diseases. In this connection, recently various ethnobotanical studies have been reported to expose the knowledge from the various tribals of Tamil Nadu (Eluvakkal, 1991; Alagesaboopathi *et al.,* 1999; Sankarasivaraman, 2000; Venkatesan, 2005; Ignacimuthu *et al.,* 2006; Sandhya *et al.,* 2006; Ayyanar *et al.,* 2008; Kottaimuthu, 2008; Shanmugam, 2008; Sukumaran *et al.,* 2008 and Shanmugam *et al.,* 2009). However, there are lot of studies available for the ethnobotanical knowledge of the scared groves, the floristic study and their ethnobotanical uses are yet to be explored from the habitat of pond storage area. Pond vegetation requires to be analysed, since pond ecosystem is unique one with the dynamic nature of its ecological components and the sediment quality varies with the different nature of the pond, eventually altering the soil quality and the associated occurrence of vegetation. The objective of the present study is to undertake the inventory on the vegetation analysis in the Vettangudi birds sanctuary pond of Sivaganga District of southern Tamil Nadu.

Materials and Methods

Locality and Description of the Experimental Ponds

The present study was carried out in Vettangudi water birds sanctuary of southern Tamil Nadu. Morphometric details of the pond were already described (Kannan and Arun Raja, 2010). VDGI pond was declared as National Wildlife Reserve in June 1977, since then this pond serves as an eco-tourist spot. Thousands of birds migrating from various parts of the country and continents to this pond for a season between the months of November and February comprised of 20 different bird species (Karthick and Banumathi, 2009).

Vegetation analysis was carried out in the ponds, followed drying in the month of September, 2009. Sampling was done using quadrat method by laying down the 1m^2 steel bars at 10 different points of randomly chosen areas inside the water storing region of the experimental ponds in the dry season. Specimens of the plants were collected and identified taxonomically with the help of Gamble and Fisher (1915-1935).

Results and Discussion

VDGI pond harbored a total of 31 plant species which are included in 19 families. Euphorbiaceae family has 4 species; Amaranthaceae and Poaceae had 3 species

Table 26.1: Vegetation analysis of Vettangudi pond habitat and the medicinal uses of the plants.

Sl.No.	Name of the Plant	Family	Common Name	Part(s) Used	Medicinal Uses
1.	*Abutilon indicum* (L). Sweet	Malvaceae	Thuthi	Seed, Root	Black patches, Ulcers
2.	*Achyranthes aspera* L.	Amaranthaceae	Nayurivi	Leaf, Root	Scorpion bite, Scabies
3.	*Alternanthera sessilis* (L). R.Br.ex Dc	Amaranthaceae	Ponnanganni keerai	Leaves	Ulcers
4.	*Amaranthus spinosus* L.	Amaranthaceae	Mullikeerai	Leaves, Seeds	Stomach ache
5.	*Calotropis gigantea* (L). R.Br.	Asclepiadaceae	Erukku	Latex, flower, root	Wound healing, fever, cough
6.	*Canthium parviflorum* Lam.	Rubiaceae	Nallakkarai	Leaves	Dysentery
7.	*Cardiospermum helicacabum* L.	Sapindaceae	Mudakkathan	Whole plant	Rheumatism
8.	*Cassia tora* L.	Caesalpiniaceae	Usithakarai	Leaves	Ringworm, ulcers
9.	*Chloris barbata* Sw	Poaceae			
10.	*Cleome gynandra* L.	Capparidaceae			
11.	*Coccinia grandis* (L) J. Voigt	Cucurbitaceae			
12.	*Croton bonplandianus* Billon	Euphorbiaceae	Eliamanakku	Leaf extract	Fever
13.	*Cyperus murex* L.	Cyperaceae			
14.	*Cyperus rotundus* L.	Cyperaceae	Korai	Tuberous roots	Stomach and bowel complaints
15.	*Datura metal* L.	Solanaceae	Oomathai	Leaves	Mental disorder, skin diseases, piles
16.	*Solanum nigrum* L.	Solanaceae	Manatakkali	Fruits	Cough, abdominal diseases
17.	*Eragrostis unioloides* (Retz).	Poaceae			
18.	*Euphorbia hirta* L.	Euphorbiaceae	Amampatchaiarisi	Leaves, latex	Venereal disease
19.	*Evolvulus alsinoides* (L). L.	Convolvulaceae	Vishnukarandi	Leaf decoction	Prolonged fever
20.	*Heliotropium indicum* L.	Boraginaceae			
21.	*Ipomoea aquatica* Forsskal	Convolvulaceae	Kalmi	Whole plant	Jaundice

Contd...

Table 26.1–*Contd...*

Sl.No.	Name of the Plant	Family	Common Name	Part(s) Used	Medicinal Uses
22.	*Mollugo cerviana* Ser.	Aizoaceae	Pappadai	Whole plant	Cold and cough
23.	*Ocimum canum* Sims	Lamiaceae	Nayitulasi	Leaves	Skin diseases
24.	*Panicum repens* L.	Poaceae			
25.	*Parthenium hysterophorus* L.	Asteraceae	Mookkithippoochedi	Inflorescence	Asthma
26.	*Pergularia daemia* (Forsskal) Chiov.	Asclepiadaceae	Velipparuthi	Leaves, Plant extract	Asthma Uterine Menstrual trouble
27.	*Pedalium murax* L.	Zygophyllaceae	Anai Nerinjii	Plant mucilage	Stomach pain, ulcers
28.	*Phyllanthus amarus* Schum and Thonn	Euphorbiaceae	Keelanelli	Root	Jaundice
29.	*Phyllanthus maderaspatensis* L.	Euphorbiaceae	Melanelli	Leaves	Head-ache
30.	*Prosopis julifera* (Sw). DC.	Mimosaceae	Karuvelam	Young pods	Diarrhoea
31.	*Acanthospermum hispidum* DC.	Asteraceae			

each; Asclepidacae, Cyperaceae, Solanaceae, Convolvulaceae and Asteraceae had 2 species each and the other 11 families were observed with the existence of single species in the dry ponds. Thirteen species occurred in less than 5 quadrats. *Acanthospermum hispidum* was found as the dominant species of the study area. Almost all the plants have been used by the local community in treating the ailments for which the details are depicted in the Table 26.1.

It is also noted in the present work that most of the studied plants' life forms is the week stemmed herbs; which shows the highly protected habitat nature, since the pond becomes as the bird sanctuary area.

Vegetation biodiversity has found a firm place for its application for varying uses, especially as healing plants, since most of the herbs have been used in the traditional healing system. Pond water quality changes and pond degradation are the conditions causing concern on such medicinal plants diversity and the livelihood depends upon them. Habitat protection makes such floristic composition intact and promoting the ecosystem restoration. Ponds with varying utilization and management make the pond ecosystem components and its functioning. Therefore, the study of the floristic composition of the different pond sources with varying utilization unfurls the details of the ecosystem components from where, livelihood and the environment functioning fully depend.

References

Alagesaboopathi, C., Dwarakan, P. and Balu, S., 1999. Plants used as medicine by tribals of Shevaroy hills, Tamil Nadu. *J. Eco. Tax. Bot.*, 23(2): 391–393.

Arora, R.K., 1997. Ethnobotany and its role in the conservation and use of Plant genetic resources in India. *Ethnobotany* 9: 6–15.

Ayyanar, M., Sankarasivaraman, K. and Ignacimuthu, S., 2008. Traditional herbal medicines used for the treatment of diabetes among two major tribal groups in South Tamil Nadu, India. *Ethnobotanical Leaflets,* 12: 276–280.

Eluvakkal, T., 1991. Ethnobotanical studies of Paliyans of Sirumalai Hills, Sathuragiri Hills and Kumily Hills. *M. Phil. Dissertation*, Thiagarajar College (Autonomous), Madurai.

Ignacimuthu, S., Ayyanar, M. and Sankara Sivaraman, K., 2006. Ethnobotanical investigations among tribes in Madurai District of Tamil Nadu (India). *J. Ethnobiol. Ethnomed.*, 2: 25.

Jain, S.K., 1964. The role of botanist in folklore research. *Folklore,* 5: 145–150.

Jain, S.K., 1991. *Dictionary of Indian Folk Medicine and Ethnobotany.* Deep Publications, New Delhi, pp. 311.

Kannan, D. and Arun Raja, T., 2010. Comparative spatial and temporal analysis of ecology of ponds with varying management practices. In: *Proceedings, BALWOIS–2010.* Fourth International Conference on Water Observation and Information System for Decision Support, Ohrid, Macedonia, 25–29, May.

Karthick, A. and Banumathi, V., 2009. Wildlife study at Vettangudi Birds' Sanctuary. *B.Sc. Dissertation*, Thiagarajar College, Madurai, 10pp.

Kottaimuthu, R., 2008. Ethnobotany of the Valaiyans of Karandamalai, Dindigul District, Tamil Nadu, India. *Ethnobotanical Leaflets,* 12: 195–203.

Parekh, J., Jadeja, D. and Chanda, S., 2005. Efficiency of aqueous and methanol extracts of some medicinal plants for potential antibacterial activity. *Turkish Journal of Biology,* 29: 203–210.

Sandhya, B., Thomas, S., Isabel, W. and Shenbagarathai, R., 2006. Ethnomedicinal plants used by the Valaiyan community of Piranmalai hills (Reserved forest), Tamil Nadu, India: A pilot study. *African J. Trad. CAM,* 3: 101–114.

Sankarasivaraman, K., 2000. Ethnobotanical wealth of Paliyar tribe in Tamil Nadu. *Ph.D., Thesis*, Manomanium Sundaranar University, Tirunelveli.

Shanmugam, S., 2008. Flora of angiosperms and ethnobotanical studies on Paliyar tribes of Pachalur in Dindigul district of Tamil Nadu, India. *M.Phil. Dissertation*, Thiagarajar College (Autonomous), Madurai.

Shanmugam, S., Anburaja, V., Muthupandi, T. and Rajendran, K., 2009. Ethnomedicinal survey on Paliyar tribes of Agamalai in Theni district of Tamil Nadu. In: *National Seminar on Recent Trends in the Coservation and Utilization of Under Utilized Wild Edible Plants*, Bharathiar University, Coimbatore, pp. 20.

Shanmugam, S., Ramar, S., Ragavendhar, K., Ramanathan, R. and Rajendran, K., 2008. Plants used as medicine by Paliyar tribes of Shenbagathope in Virudhunagar district of Tamil Nadu. *J. Eco. Tax. Bot.,* 32 (4): 922–929.

Sukumaran, S., Jeeva, S., Raj, A.D.S. and Kannan, D., 2008. Floristic diversity, conservation status and economic value of miniature sacred groves in Kanyakumari District, Tamil Nadu, Southern Peninsular India. *Turk. J. Bot.,* 32: 185–199.

Singh, H.B., 1997. Alternate source for some conventional drug plants of India. In: *Ethnobotany and Medicinal Plants of Indian Subcontinent* (2000), (Ed.) J.K. Maheshwari. Scientific Publishers, Jodhpur, India, p. 109–114.

Venkatesan, G., Ganesan, S., Kesavan, L. and Banumathy, N., 2005. Ethnobotany of ethnic group Thottianaickan of Tiruchirapalli district, Tamil Nadu, India. In: *Advances in Medicinal Plants*, Vol. 1. Asian Medicinal Plants and Health Care Trust, Jodhpur, pp. 59–75.

2013, Glimpses of Animal Biodiversity   *Pages* **233–240**
Editor: **Dr. K. Muthuchelian,** *Vice Chancellor, Periyar University, Salem*
Published by: **Daya Publishing House, NEW DELHI**

Chapter 27

Role of Bioinformatics in Biodiversity Data Management

Piramanayagam Shanmughavel

Department of Bioinformatics,
Bharathiar University, Coimbatore – 641 046

ABSTRACT

Concomitant with the growing interest world-wide in the conservation and sustainable management of biodiversity of animals, there has been a four-fold increase in the volume of literature published during the past two decades. Recording and dissemination of this information were almost exclusively using printed publications. Development of bioinformatics databases have lead to increasing proportion of this pool of information now being held in databases which can either be searched on line from remote sites. The role of bioinformatics in biodiversity data management and integration of scattered data is discussed. Further, how to create a digitized database is also narrated.

Level of Biodiversity

India, with 2.4 per cent of the world's area, has over 8 per cent of the world's total biodiversity, making it one of the 12 mega diversity countries in the world. This status is based on the species richness and levels of endemism recorded in a wide range of taxa of both plants and animals. This diversity can be attributed to the vast variety of landforms and climates, resulting in habitats ranging from tropical to temperate and from alpine to desert. Adding to this is a very high diversity of human-influenced ecosystems, including agricultural and pasture lands, and a diversity of domesticated plants and animals, one of the world's largest. India is also considered one of the

---

* Corresponding Author: E-mail: shanmughavel@buc.edu.in

world's eight centers of origin of cultivated plants. Being a predominantly agricultural country, India also has a mix of wild and cultivated habitats, giving rise to very specialized biodiversity, which is specific to the confluence of two or more habitats. About 45,000 to 47,000 plant species are reported to occur in India, representing 11 per cent of the known world flora. Nearly 90,000 species of fauna have been reported, altitude over 7 per cent of the worlds reported animal diversity. There exists considerable information on the patterns of species richness, endemism and the diversity of different plant groups (angiosperm, gymnosperms, pteridophytes, lichens, bryophytes, algae, fungi), various animal groups (including marine and terrestrial), and micro-organisms, but much more needs to be understood and appreciated. Information on micro-organisms is especially deficient [1].

Biodiversity Informatics

These new applications belong to the emerging field that we can term Bioinformatics. As the word bioinformatics is now applied universally to genomics and proteomics applications, a new term may be needed to describe applications at the organism level. Biodiversity Informatics then includes the application of information technologies to the management, algorithmic exploration, analysis and interpretation of primary data regarding life, particularly at the species level of organization. Biodiversity Informatics analyses and applications are characterized by a number of novel features. Investigators are increasingly able to use large quantities of biodiversity data—that is, analyses are now frequently based on records numbering 104 or more [2].

Bioinformatics in Biodiversity Data Management

The first step in biodiversity conservation is documentation based on the availability of information about each species with all details starting from its systematic position to the latest molecular aspects. The variability in biological data is quite natural because individuals and species vary tremendously [2]. Organism of the living world is unique in itself. This uniqueness of individuals is the diversity in life forms, which is the most important aspect of the living world. Different places in different parts of the world have their own typical kinds of living beings. This means that the extent of diversity is boundless [3].

Biodiversity Assessment

There are several stages in the process of handling biodiversity data, applicable to all level of biodiversity information: Location and acquisition of material, measurement and recording of data analysis summarization and interpretation of results (whereby data is transformed into information), Documentation and dissemination Integration (whereby information is transformed into knowledge) [1].

Measurement of Biodiversity

Biodiversity information can be summarized in Species checklists (for example in geographical area of interest) Distribution maps (of species of interest; see for example BONAP), Species diversity maps (showing "hot spots" etc.). All these can be generated from databases. In assessing the factors affecting biodiversity in an area, investigation of models and testing of hypothesis is necessary [2].

Database Creation

The challenge before us is to set standards and make technological choices that would acilitate networking of databases, and add real value to the information being brought together, while at the same time, (a) maintain the autonomy of the various databases and ensure that there is abundant scope for expression of creativity and originality of the designers and managers of different databases, as also (b) ensure the security of the data and (c) protect all legitimate intellectual property rights. This would obviously have to be worked out as a group exercise by all concerned institutions and individuals. A first step would have to involve (a) an inventory of the on-going Indian efforts, and (b) a review of the various pertinent standards, technologies and protocols developed anywhere in the world. These surveys would have to address issues of data (i) characterization and classification, (ii) validation and authentication, (iii) organization and structuring, (iv) storage, archival, warehousing, (v) retrieval, (vi) dissemination, (vii) sharing and interoperability, (viii) access, (ix) security, and (x) visualization, analysis and value addition, as well as (xi) use of multiple languages, and (xii) capacity building needs.

Entities

While we did not have adequate time at the Workshop to deal with the large number of issues that would have to be addressed, we would like to offer a few preliminary suggestions. Firstly, the scope of the databases that need to be thus brought together would have to go beyond the taxon-centric database that is the exclusive focus of many efforts including the Global Biodiversity Information Facility (GBIF). Thus we visualize linking of data on medicinal uses and chemical composition to taxonomic data on medicinal plants. We also need to link taxonomic data on medicinal plants to data on geographical distribution, abundance, and harvest levels on land under different forms of ownership and access regulations to support development of strategies for conservation and sustainable use of these plant populations. We therefore suggest that the networking effort brings under its purview databases that will deal with the whole range of categories of entities listed in Table 27.1.

Architecture

Of course, there are many useful lessons from the on-going international exercises for us. Thus, the International System of Patent Classification might serve as a useful starting point for a system of classification of database entities. One may particularly mention here the Traditional Knowledge Resource Classification being developed as a component of this system as a result of Indian inputs. We might also with profit build upon the Data Architectural Model of the Global Biodiversity Information Facility (GBIF). GBIF intends to make world's biodiversity information available to all within the next 10 year period. Employing the architectural model summarized in Table 27.2, the GBIF is currently serving over 10 million records from 34 distributed databases.

Table 27.1: A possible framework for definition of entities for Biodiversity Databases.

Class of Entities	Sub-class	Examples
Geographic	Physical	Mountain ranges, river basins, wetlands
	Political	Local bodies, states, UTs, GoI
	Ecological	Agro-ecological zones, Biomes, Landscape elements
	Management regimes	Reserved forests, wild life sanctuaries, sacred groves
Biological	Genes	Bt genes in Bt cotton
Units	Taxonomic groups	Species, orders
	Natural/cultured	Varieties registered under Protection of Plant Varieties and Farmers' Rights Act
	Management status	Species listed under Schedules of Wild Life Act, CITES, or in Red Data Books, sacred species
People	Political units	Citizens of India, citizens of particular states or Panchayats,
	Membership of organizations	University faculties, Staff of R&D labs
	Occupational groups	Hakims, Vaids, Biotechnologists
	Social groups	Indian communities as recognized by Anthropological Survey, Scheduled tribes
	Holders of material property rights	Holders of rights of collection of forest produce from particular localities
	Holders of intellectual property rights	Holders of patents
	Linguistic units	Speakers of different languages and dialects
Organizations	Relation to government	Government agencies, NGOs, Inter-governmental agencies
	Objective	Commercial, Not-for-profit, Scientific
	Nationality	Indian, foreign
Biological Populations	Population levels	Population levels of particular species in specific localities
Manipulations		Harvests or production under cultivation, including specific techniques employed, of particular produce of particular species in specific localities
Transport		Transport of produce of particular species in specific localities to other specific localities
Marketing		Marketing of produce of particular species in specific localities in other specific localities
Biological Materials	Processing	Preparation of acetone extracts from particular plant species
	Value addition	Preparation of plant based drugs or cosmetics
	Technology	Stabilization of alkaloids extracted from Neem
	End products	Particular molecules isolated from biological sources
	Uses/disservices	Therapeutic, cosmetic, allergenic

Contd...

Table 27.1–*Contd...*

Class of Entities	Sub-class	Examples
Knowledge	Nature of knowledge	Satellite imageries, folk taxonomies, medicinal properties of particular species
	Source of knowledge	Scientific research, Classical tradition, Individual hakim
	Medium	Scientific journals, palm-leaf manuscripts, oral traditions
	Generation of knowledge	New collaborative research, new Indian research, grass-roots innovations
	Transmission of knowledge	Web-sites, scientific publications, scientific meetings, orally from mother to daughter
	Access to knowledge	Public domain, patented, trade secrets
	Rights over knowledge	Individual, community, corporate
	Validation	Raw, Validated through different techniques
Disputes	Disputes over access to material or knowledge resources	Refusal of permission to an application for collaborative research

Table 27.2: GBIF Web Services Architecture

Registry services	UDDI
Interface description	WSDL
Access protocols	SOAP, DiGIR
Data encoding	XML, XML Schema
Transport	HTTP over TCP/IP

Data Quality

Data quality is clearly an overriding concern. A whole series of standards will have to be developed and checks organized at the level of data creation, organization and sharing to ensure high quality. Here one may mention a modern tool that may be employed in connection with data meant to be publicly available, namely, *Wiki-wiki* pages. These web-pages permit any visitor to edit portions designated as open to editing. This permits all interested parties to correct mistakes and add to the information. Of course it is possible that some visitors may maliciously distort the material. However, this is not a serious problem since the earlier versions can be preserved and reinstated in case of such a mischief. The experience of *Wikipedia,* a public knowledge resource encyclopedia has been very positive. Portions of the biodiversity database may therefore be thus maintained and further developed as *Wiki* pages.

Outline of the Technical Data Collection Format

Technical contact person from the organization with contact details

Details of the Database

☆ Name of the database

☆ Objectives and aims of the database

☆ Brief description

Mechanism Adopted in Data Compilation and Digitization
Details of data (primary/secondary)

Data source

☆ Indicate the mechanism adopted in data validation -

☆ Stakeholders of the database

☆ Country, institutions, individuals providing the data

☆ Other sources of data

☆ Metadata details

Table 27.3: Kinds of entities represented in the database.

Class of entities	*Sub-classes*
Geographical	Physical, Political, Ecological, Management regimes
Biological units	Genes, Taxonomic groups, Natural/cultured, Management status,
People	Political units, Membership of organizations, Occupational groups, Social groups, Holders of material property rights, Holders of intellectual property rights, Linguistic units
Organizations	Relation to government, Objective, Nationality
Biological populations	Population levels, Manipulations, Transport, Marketing
Biological materials	Processing, Value addition, Technology, End products, Uses/ disservices
Knowledge	Nature of knowledge, Source of knowledge, Medium, Generation of knowledge, Transmission of knowledge, Access to knowledge, Rights over knowledge, Validation
Disputes	Disputes over access to material resources, Disputes over access to knowledge

1. Priorities addressed by the database (taxonomic, political, conservation, policy making, socio-economic)
2. Targeted user groups (policy makers, scientific community, educators, private company, general public, students)
3. Contractual agreements in place or desired for data sharing/output/ acquisition (formal contract, MOU, letter of agreement, verbal contract)
4. How IPR, Copyright and financial gain issues are handled source of funding
5. Total cost incurred on database so far and projected total cost to complete the database
6. Problems and hurdles faced/expected

Technical Specifications

1. Platform and Technology used (Windows/Linux, MySQL/Oracle/XML) with version of each

2. It is already web enabled in some format?
3. Data types incorporated (text, dbf, mdb, xls, multimedia formats)
4. Data transfer methods (CD/Print Output/Email/Internet/Interactive web search)
5. Details of the database like Entities, Relationships, Structure, Fields, etc.

Number of Data Tables

Enclose E-R Diagram and Tables with relationships

1. Method of data capture and data entry *e.g.*
   a. If data captured by field workers, if validated by others and then entered
   b. Accepted from literature
   c. Entered by hand-held gadgets/through user interface etc.
2. Current access mechanisms
3. Size of database, number of records, field densities
4. If data has temporal dimension *i.e.* data may change with time. If so whether old data is preserved with time tag.
5. Procedure followed in editing data, who is authorized, if pre-edited copy is retained etc.
6. Code sets used in storing database
7. Presence of multilingual component and details
8. Standards being followed
9. Time scale for developing the database and completion of the database
10. Estimated cost and time/record
11. Special Remarks

Feasibility and Willingness to Participate in "Network of Indian Biodiversity Databases"

1. Exchange/sharing data is feasible
2. Exchange formats/standards adopted (xml, z39.50, CORBA)
3. Level of access restrictions *i.e.* if access restrictions are imposed then its level : table/record/field
4. Details and contact information of other related databases you are collaborating with list of key publications, reports and websites resulting from Database.

Users of Biodiversity Information

The users for biodiversity information depend on who the users are. Mainly this information is used in the form of models, hypothesis tests, summaries, reports, classifications and checklists.

Primary users are: Scientist those are working in the same field. Managers (of reserves) Legislators Checklists (of endangered species) Agriculturalists Information on species or varieties suitable for breeding programmers, models (for prediction) on pest controllers Pharmaceutical companies [4].

Importance of Biodiversity Database

Conservation of threatening species for future generation includes making them as a specimen, images and etc. finally organizing them into database (Khoshoo1995). Though the Bioinformatics discipline was evolved for the management of genomic and proteomic data recently it is encroached in all areas of biology and concedes that it is not a supporting branch of science but a subject that direct future course of research in life sciences. In the past decade electronic storage media, world wide web and internet, database technology and in digitalization of data creation of public access databases are creating a revolution in the way that biodiversity information is created maintained, distributed and used. Biodiversity informatics then includes the application of information technologies to the management, algorithmic exploration, analysis and interpretation of primary data regarding life particularly the species level of organization. In the internet hundreds of URLs are providing biodiversity informatics by various organizations. Trees are important source of food. Leaves, seeds, and fruits give rich variety of edible material. Presently here is a revival of interest in re-examining the traditional uses of native trees including the medicinal, edible fruits, nuts and other resources. There is little or no effect for genetic improvement to enhance the production of timber, firewood, fodder, fruits etc. Any products from trees can be achieved only if there is a basic understanding or knowledge about genetic variability, breeding systems, penology, pollination biology, seed production, seed viability, dormancy, germination, establishment and several other aspects like ecology, studies related to R and D output. Such aspects are not available for an even single useful tree species. The importance of conserving the biodiversity of native origin for the sustainable use on a long run needs details regarding theses species and also many of them are endemic species which has got multiple uses such as timber, firewood, fodder etc. based on the information and sources they can be into several databases [5].

References

1. F.A. Bisby, Biodiversity informatics and Internet. Science 283:2309 (2000). [PMID: 11009408].

2. R. K. Colwell and J. A Coddington, Phil. Trans. R. Soc. 335: 101–118 (1994) PMID: 7972351.

3. W.Jetz, and C.Rahbek, Science 297: 1548–1551 (2002) PMID: 12202829.

4. J Soberson and A.Townsend Peterson, *Phil. Trans. R. Soc. Lond* 359:689-698 (2004) PMID: 15253354.

5. C. Rahbek, and G. R. Graves,. Proc. R. Soc. Lond. B 267: 2259–2265 (2000) PMID: 11413641.

Index